"Business Ecology provides an ins_____ _____ _____ __ ____ __ build strategic relationships with your customers, employees, suppliers, local communities, and other stakeholders. I highly recommend this book to leaders who wish to build community within and outside their organizations."

—**Bryan M. Thomlison,** Thomlison Strategic Alliances, Inc.

"Business Ecology is essential for leaders and managers who see that business plays an indispensable role in creating a sustainable future . . . one that builds on economic vitality, social well-being, and protection and restoration of the environment. This book sets out in useful, operational detail the emerging organic models of the twenty-first century business organization. These models combine integrity instilled by vision and strong core values with the flexibility and street smarts needed for competitive success."

—**Robert L. Olson,** Research Director, Institute for Alternative Futures

"All signs point to nature-based business strategies as the basis of a sustainable future. *Business Ecology: Giving Your Organization the Natural Edge* is an eloquently written, comprehensive and practical guide for managers who want to act upon this potent source of innovation."

—Jacquelyn A. Ottman, President, J. Ottman Consulting, Inc.;
author, *Green Marketing: Opportunity for Innovation*

"The authors are successful in formulating an ecological model for organizational management and design, synthesizing the thinking of visionaries and leaders from a variety of fields. In doing so, they have enlarged the scope of the traditional business paradigm with the wisdom of placing human and environmental values at its center. The authors provide structure, language, and principles which, when combined with the variety of examples from natural systems, define the way businesses will operate in the future. I highly recommend *Business Ecology* to anyone who is interested in helping to achieve a sustainable future, and especially to the managers and leaders who see the vital role of business in creating one."

—**John F. Elter,** Vice President, New Business Development and Chief Engineer,
Office Document Products Group, Xerox Corporation;
Co-Founder of the nonprofit, ESSE, Engineers and
Scientists for a Sustainable Environment

"Finally—a book with a vision for a sustainable future and practical strategies for making it happen. *Business Ecology* distills traditional perspectives on profitability and growth with broader, deeper, interconnected goals, such as creating shared values, community, personal growth, and purposeful work. A wealth of examples from visionary organizations, such as Monsanto, Mitsubishi Electric America, Businesses for Social Responsibility, and Chez Panisse makes the case for a values shift towards sustainability."

—**Claudine C. Schneider**, Director, Land and Water Fund of the Rockies; former US Congresswoman, Rhode Island

"This has some of the best examples I've seen of how to follow nature's blueprint to be sustainable, competitive, and in sync with ecology and economy. Corporate planners, managers, and designers—anyone who still thinks that the environment and economy are not intimately related—will find this book a useful tool. *Business Ecology* is a redesign of the way we see ourselves and our organizations. It is a needed change."

—**Carl L. Henn**, Senior Vice President, Royal Capital, Inc.; Adjunct Lecturer on Industrial Ecology, Rutgers University

"The more we understand how we're connected, the more we see how we truly belong. The same principle applies to business. Be sure to have a fine-point marker at the ready when you read this connection-packed book! You'll be wanting to make notes and cross-references on every page!"

—**Martha I. Finney**, co-author, *Find Your Calling, Love Your Life*

BUSINESS ECOLOGY

BUSINESS ECOLOGY

Giving Your Organization the Natural Edge

JOSEPH M. ABE
PATRICIA E. DEMPSEY
DAVID A. BASSETT

Butterworth–Heinemann

Boston Oxford Johannesburg Melbourne New Delhi Singapore

Butterworth–Heinemann supports the efforts of American Forests and the
Global ReLeaf program in its campaign for the betterment of trees,
forests, and our environment.

Library of Congress Cataloging-in-Publication Data
Abe, Joseph M. (Joseph Michael), 1958-
    Business ecology : giving your organization the natural edge / Joseph M. Abe,
Patricia E. Dempsey, and David A. Bassett.
        p.   cm.
    Includes bibliographical references and index.
    ISBN 0-7506-9955-8 (pbk. : alk. paper)
    1. Industrial organization. 2. Industrial management.   I. Dempsey, Patricia E.
(Patricia Elizabeth), 1957–   II. Bassett, David A. (David Allen), 1949–   III. Title.
HD31.A223 1998
658—dc21   97-46458
                                                                                                                CIP

British Library Cataloguing-in-Publication Data
A catalogue record for this book is available from the British Library.

The publisher offers special discounts on bulk orders of this book.
For information, please contact:

Manager of Special Sales
Butterworth–Heinemann
225 Wildwood Avenue
Woburn, MA 01801-2041
Tel: 781-904-2500
Fax: 781-904-2620

For information on all Butterworth–Heinemann books available, contact our
World Wide Web home page at: http://www.bh.com

10 9 8 7 6 5 4 3 2 1

Printed in the United States of America

Contents

List of Figures

List of Tables

About the Authors

Joseph M. Abe is co-founder and president of the Business Ecology Network (BEN), a nonprofit learning community for leaders and managers who want to apply a new way of thinking—business ecology—to create new, sustainable opportunities for their business, government, and nonprofit organizations. BEN's mission is to be a catalyst for life-sustaining enterprise. He is also principal of Business Ecology Associates, a for-profit consulting group that helps businesses and other organizations apply business ecology to develop business-based solutions that integrate profitability, values-based management, stakeholder relations, life-cycle thinking, and environmental performance. Mr. Abe formerly cofounded the U.S. Environmental Protection Agency (EPA) Futures Program, which the Carnegie Commission on Science, Technology, and Government recognized as a model for long-range planning. He has diverse experience and education related to strategic planning, environmental policy, industrial ecology, sustainable development, organizational learning, and pollution prevention. His professional interactions have included the President's Council on Sustainable Development, the Global Business Network, the EPA's Science Advisory Board, and the White House Office of Science and Technology.

Patricia E. Dempsey, co-founder and communications director of the Business Ecology Network, is a poet, essayist, and editor with a special interest in enhancing community and economic development by profiling local culture, business, food, and the arts for such publications as the *Washington Post*. Recent projects include documenting the renaissance of local beach towns and oral histories from traditional waterman communities around the Chesapeake Bay. A former co-founder of The Idea Consortium, an educational nonprofit, she has expertise in creating stakeholder information programs for small businesses and nonprofit organizations as well as a background in visual design and social history/anthropology. She is raising her two sons, Alex and James Kriz, and pursuing an M.A. in writing at Johns Hopkins University.

David A. Bassett is a founding director of the Business Ecology Network. He helped make the phrase "Pollution Prevention" a buzzword of the 1990s, as a former member of the Office of Pollution Prevention at the U.S. Environmental Protection Agency. In this position, he developed a Pollution Prevention Strategy for Energy and Transportation as a road map for integrating environmental and energy-efficiency goals. He also cofounded the National Industrial Competitiveness through Energy, Environment, and Economics (NICE^3) Program. As a member of a National Performance Review Team, he was a voice for comprehensively rethinking the nation's approach to environmental protection. He was the first director of energy programs for Erie County, New York, and helped define the Interstate Air Quality Control Regions for the Commonwealth of Kentucky. He has written a number of technical papers and currently supports several working groups of the President's Council on Sustainable Development. He is an advocate for innovation, cleaner technologies, and native peoples.

Preface

Nature is always hinting at us. It hints over and over again.
And suddenly we take the hint.

Robert Frost

Turning to nature for reflection, inspiration, and solutions to the complexities of life that frustrate and puzzle us is nothing new. For centuries, poets have been inspired by it, architects have emulated its engineering feats, and scientists have marveled over its mysteries, championing its elegant solutions. Even writers and philosophers have shared insights drawn from their simple observations of nature.

The writer Anne Lindbergh, for instance, author of *Gift From the Sea,* reflects on how the design of various seashells represents the changing stages of human life and relationships. She draws her inspiration from the sea, its gifts, and the peaceful solitude of beach living. When Lindbergh focuses on individuals and the need to be self-referring to a solid "center" of core values, she states:

> Have we been successful, working at the periphery of the circle and not at the center? . . . If we stop to think about it, are not the real casualties in modern life just these centers I have been discussing: the here, the now, the individual and his relationships. . . . [W]hen we start at the center of ourselves, we find something worthwhile extending toward the periphery of the circle. We find again some of the joy in the now, the peace in the here, some of the love in me and thee which go to make up the kingdom of heaven and earth.

So it is with businesses and organizations. We must start at the center and establish core values to shape actions and vision. It is in this spirit that we present business ecology, a new field for sustainable organizational management and design, that is based on the principle that organizations, as living organisms, are most successful when their development and behavior are aligned with their core purpose and values—what we call "social DNA." This social DNA creates a solid inner, self-referential circle upon which the rest of the business or organization can keep extending to include all aspects of its viability, such as product design, manufacturing, stakeholder relations, marketing, and ecological efficiency.

We wrote this book for leaders and managers who see the vital role of business in creating a sustainable future and who are in positions to make a difference. It is intended for those leaders who recognize our enormous environmental, economic, and social challenges, as well as the promise of integrating profitability, values-based management, stakeholder relations, and environmental performance. We offer a new way of thinking—business ecology—that we are convinced will create new, transforming opportunities for their business, government, and nonprofit organizations. Business ecology does this in several ways: it takes a closer look at natural systems, uses systems thinking, builds strong relationships, creates ecological efficiency, develops closed-loop behavior and multiple value-creation cycles, and merges profitability and stakeholder relations with a commitment to the local environment and community as well as a larger, global web of resources and relationships—one of the outermost circles that surrounds us all. Business ecology's mindware—organic models for organizational management and design—synthesizes innovative thinking from leaders and visionaries in fields such as management, biology, philosophy, the arts, public policy, along with profiles and case studies of sustainable enterprises and ecological thinking from businesses and nonprofit organizations. Some of these include: VISA, Xerox, Malden Mills, Monsanto, Interface, Growth Cycle Design, Inc.; aquaculture; business ecosystems such as Kalundborg, Denmark and Cape Charles, Virginia; and organizations such as Sustainable Technologies Corporation, Chez Panisse, Vermont Businesses for Social Responsibility, and The Natural Step.

We believe that those leaders who are able to grasp the implications of sustainable development and transform their businesses and organizations into sustainable enterprises will have a clear, competitive advantage in the new, emerging ecological economy. Unlike the linear, extractive industrial-age economy, the ecological economy is cyclical, symbiotic, and resource efficient. Business ecology takes the first step of defining a common language that links our industrial past with the ecological future. It emulates natural systems design and provides an elegant, relationship-oriented approach that leaders, employees, and other stakeholders can use to see how their organizations really work.

In researching this book, we discovered over and over again that increased efficiency and growth can be created by applying—among other things—the cyclical thinking that already exists throughout natural systems. The life cycle of value creation, for instance, is an inner circle, a part of the larger design of your business or organization. Successful value creation stems from an organizational design that clearly defines its underlying core purpose and values. This "social DNA" is deeply woven into our worldview and reflects the passing down, through the centuries, of beliefs and priorities borrowed from traditional and ancient cultures. Some of the wisdom from these societies is particularly relevant today, including a respect for natural rhythms, collective cooperation, and the strong organizing forces of belief systems, shared rituals, and traditions. In the new ecological economy, viable, adaptive organizations will adopt such traditional wisdom along with modern organizational models that emulate natural systems.

In addition, by applying nature's holistic organizing elegance, leaders and managers can measure overall (not just financial) viability to see and renew the vital relationships that sustain their business or organization. Business ecology spotlights those intangible elements of a business or organization's design, such as stakeholder relations, core purpose and values, community, value-creation cycles, and innovative thinking, that are essential factors shaping its future success. Business ecology is a powerful catalyst for integrating economic, social, and environmental goals, and a natural strategy for thriving in the next economy.

Life is a journey to be savored and shared. We offer this book as a discovery guide into the world of purposeful, life-sustaining enterprise. We hope it provides useful insights for those organizations and business leaders who dare to bring about organizational change. There is a hunger for change, for a new paradigm. People are taking back their lives where they work and live, acting as catalysts for positive growth and rejuvenation. Our intent is to plant strategic seeds. Seeds of hope, inspiration, and opportunity for those who have dedicated their lives to creating a more livable, just, tolerant, and sustainable society.

There are signs of hope that humankind is evolving to meet the many challenges we face: visionary business leaders who are establishing higher levels of success, corporate responsibility, and accountability; innumerable grassroots programs to revitalize communities; efforts to reconnect people with their political system; strong, broad-based commitments to environmental protection, social justice, and economic opportunity; and a reconnection to common sense and basic values. However, bringing about these changes depends largely on healthy, effective, and responsive organizations, it depends on bringing complex systems into balance. Are organizations capable of comprehending and meeting these challenges? Can we learn and evolve fast enough to thrive in the next century?

To help leaders and managers of businesses and other organizations make a fluid transition from the industrial to the ecological economy, we have created the seven seeds of business ecology: (1) create a values-based organization, (2) adapt and thrive, (3) use a new language for success, (4) develop vital flows and relationships, (5) merge art and science, (6) learn from natural systems, and (7) put community back in business.

We wrote this book to help each of us break free of the old perspective and open our eyes to ways we can evolve rapidly toward a sustainable future. Meeting this challenge requires a new way of thinking. This requires a worldview shift, a fundamental transformation in how we see nature and ourselves. We must, as the biologist Rachel Carson suggests, learn to master ourselves if we are to live compatibly with the environment and each other.

Central to this evolution in human thinking and behavior is the transformation of organizations. The means for organizational transformation are sprouting up in diverse, interdisciplinary fields, such as industrial ecology, organizational learning and psychology, workplace health and ergonomics, ecological accounting, sustainable development, and ecological economics. The central objectives of this book are to synthesize these innovations into a new language for organizational success, to provide examples and metaphors that illustrate insights and lessons learned, and to be catalyst for life-sustaining enterprise.

This book is for people who care about life, the planet we inhabit, and the future we will leave our children. It is about discovering, as the writer Anne Lindbergh once did, infinite wisdom and inspiration in the world around us. Perhaps one day organizations will be designed with the grace and spiraled elegance of a seashell. Why not? It is possible with a new, sustainable way of thinking.

Joe Abe
Trish Dempsey
Dave Bassett

The Business Ecology Network
P.O. Box 29
Shady Side, Maryland 20764-9546 USA
410-867-3596

Acknowledgments

Creating a book involves many people: those who are actively involved in its research, design, and writing, and many more who provide vision, guidance, and inspiration, sometimes from across the divide that separates this world from the next. This includes the leaders, philosophers, visionaries, and organizations mentioned throughout this book; they have defined those things that are timeless and essential to us all, such as nature, relationships, and community. They are redefining "success" and leading the transformation to a sustainable society.

We are especially thankful to our families and friends who provided love, support, and understanding during the arduous task of creating and reworking the manuscript. We are also grateful to Eleanor and Jim Leonard for graciously hosting our Business Ecology Roundtables in Marshall, Virginia and to those who gave up their Saturdays to attend: Michael Mastracci, Tip Parker, Gregg Freeman, Stanley Serfling, Katherine O'Dea, and Alan Schroeder. Thanks also to Carl Henn, Thomas Gladwin, Claudine Schneider, Bryan Thomlison, Kate Fish, John Elter, Bob Olson, and Jacquie Ottman for reviewing the manuscript and providing thoughtful insights and spiritual support.

Thanks are due to Karen Speerstra, Stephanie Gelman, and Maura Kelly of Butterworth-Heinemann for their helpful comments and suggestions. A note of thanks also to Maya Porter for referring us to Karen Speerstra in the first place. We'd like to make a special mention of our parents, Al and Elizabeth Abe, Edward and Louise Dempsey, Marlin and Nellie Bassett, our children, Jonathan Bassett, Alex and James Kriz, and wife, Jean Bassett, for inspiration, guidance, and encouragement.

We are also grateful for the community of people—our friends, families, teachers, clients, mentors, artists, and visionaries—who have shaped our lives. And finally we wish to acknowledge the Creator for the magnificent beauty and organizing elegance of nature—the ultimate source of inspiration for this book and the divine wisdom and creativity within each of us.

1

Mindware for the New Millennium

One has to be dumb, deaf and blind to not see that we are in the midst of a global institutional failure. . . . Society is in the midst of a millennial change that will dwarf the industrial revolution in a tenth of the time.

Dee Hock

Most of us instinctively recognize that the old ways of running organizations, which can be traced back to the 1700s and earlier periods, cannot meet the challenges of the 21st century. The English poet Matthew Arnold once wrote of "wandering between two worlds, one dead, the other powerless to be born." Clearly we are faced with such a dilemma today.

Where is the blueprint for the future? The genetic code for organizations? Leaders and visionaries are already articulating what is a fundamental shift—from industrial age values and designs to ones shaped by biological sciences, the emerging ecological paradigm—the new ecological economy. Collins and Porras' *Built to Last*, de Geus' *The Living Company*, Wheatley's *Leadership and the New Science*, Hawken's *The Ecology of Commerce*, Moore's *Death of Competition*, Senge's *The Fifth Discipline*—the plethora of books on the subject is just the beginning. Businesses and organizations are no longer seen as hierarchical machines. Rather, they are living organisms with a web of stakeholder relationships, product cycles, and values-based genetic coding. They are flexible, responsive, and cohesive because of articulated, shared values and purposes. Their employees are connected to these values and motivated by purposeful work. Their stakeholders and communities are supportive, their products are competitive,

1

resource efficient, and their profits are steadily increasing. Companies such as Monsanto, Xerox, Interface, 3M, Malden Mills, and AT&T see the coming ecological transformation and its implications for business. They are already applying systemic, closed-loop thinking to create enduring success, competitive strength, resilient organizations, and strong shared values among employees, leaders, and managers. They are innovative, adaptive organisms. How have they embraced these changes? How can your business or organization be similarly successful as it transitions to the new ecological economy?

A NEW WAY OF DOING BUSINESS

Business ecology, a systemic, comprehensive rethinking of business, offers an answer. By emulating natural systems design, it presents mindware for the new millennium—values-based, close-looped models for organizational management, which integrate profitability, stakeholder relations, life-cycle thinking, and environmental performance. To successfully position your business or organization for the future, begin by cultivating these seven seeds of business ecology:

1. Create a value-based organization
2. Adapt and thrive
3. Use a new language for success
4. Develop vital flows and relationships
5. Merge art and science
6. Learn from natural systems
7. Put community back in business

"Nature does nothing uselessly" is a practical observation attributed to Aristotle. Nature is in fact exquisitely efficient. Business ecology melds such efficient, common sense observations of nature with systems thinking to improve your organization or company's performance in all areas. Within the framework of business ecology, for instance, businesses and organizations are viewed as organisms exchanging resources within life-supporting business ecosystems. Chapter 2 shows how this systems perspective and life-cycle thinking can transform your business or organization's efficiency and product design. Chapter 3 defines your business or organization's niche, metabolism, and habitat, including three vital processes—values-based organization and development, value creation, and cyclical flows.

Business ecology uses an ecological, holistic lens for viewing all of the relationships and flows that affect your organization. Chapter 4 looks beyond cash flow with a systemic approach to accounting that is shown to improve organizational performance, including higher revenues, profitability, new markets, lower liabilities, pollution reduction and prevention, higher motivation and productivity, and healthier stakeholder relations. Chapter 5 expands your perspective on business ecosystems, their resource exchange relationships and the factors, such as shared marketing and values, which contribute to their success. Chapter 6 examines key systemic relationships that create community within and outside of your business or organization; this includes aligning vision, values, and purpose, and building strong stakeholder relations. Shifting organizational values towards sustainability is the focus of Chapter 7; as shown by several companies, this type of organizational transformation leads to competitive advantage. To give context to these discussions, it makes sense to look at the seven seeds of business ecology and the thinking of seasoned professionals and leading visionaries that brought us to this new field of organizational management and design.

THE SEVEN SEEDS OF BUSINESS ECOLOGY

 1. Create a Values-Based Organization

Your organization, as a living organism, is most successful when its development and behavior are aligned with its vision, values, and purpose—its "social DNA"—and its environment. Business ecology, as a values-based, organic model for organizations, creates an inner core of community, continuity, and resiliency. This social DNA which defines your organization's identity, also selectively filters value-creating flows and relationships that sustain it within its environment. These flows and relationships, in turn, define your organization's internal economy (metabolism), how it makes a living (niche), and where it lives (habitat).

 2. Adapt and Thrive

Leading companies are already applying systemic, closed-loop thinking to create enduring success and competitive advantage. Your business or organization can emulate these visionary companies by applying business ecology's organic model to create a dynamic learning organization, strate-

gically positioned with vision, foresight, self-knowledge, and systems think-
ing. This model can help your organization identify and maintain its core
identity while remaining open to the changing environment. Your business
or organization can learn to adapt, innovate, survive—even thrive—in the
ecological economy.

3. Use a New Language for Success

Business ecology is the language of sustainable enterprise and the ecological
economy. This new language articulates the transition from industrial,
mechanistic thinking to biological metaphors for management and design.
It describes holistically your business or organization's well-being; aligns
its vision, values, and purpose, defines its value creation relative to all of
its stakeholders, and enhances the viability of your business or organiza-
tion's life-sustaining flows and relationships within its environment. As a
sustainable enterprise, you organization's success is defined by how effi-
ciently it optimizes value for stakeholders while enhancing and sustaining
life. In this sense, its value creation, and ultimately its success, is measured
by the extent to which quality of life genuinely improves.

4. Develop Vital Flows and Relationships

Business ecology, as an organic lens and framework, redefines organiza-
tional management, accounting, and economic development. It helps you
see and develop the vital flows and relationships needed to sustain your
business or organization within its environment. For instance, the business
ecology lens widens your organization's perspective from "cash flow" to
"life flows" and from "accountability to shareholders" to "accountability
to stakeholders." Cyclical flows—such as money, energy, materials, infor-
mation and ideas, products and services, people and other organisms, air,
food, and water—sustain economic activity at multiple levels; these include
products and services, individuals, processes, firms, communities, and
economies. Business ecology is relationship oriented. Your organization can
build constructive, closed-loop relationships with other organizations, such
as resource exchange networks in a business ecosystem, while enhancing
its own ecological efficiency, sense of community, and profitability. Busi-
ness ecology builds synergistic relationships that can reconnect your busi-
ness or organization with its community and local, regional, and global
economies and environments.

5. Merge Art and Science

Business ecology is both art and science. It incorporates the ability to see patterns, systemic relationships, different perspectives, and express creative imagination, which are traditionally the skills of an artist. It is also a science, in that it draws conclusions and creates models for organizations based on a close observation of natural systems. It integrates leading-edge thinking and proven success strategies into mindware—organic management models that save you time, money, and resources. As a synthesis, business ecology distills the best thinking from diverse fields and insights from natural systems, making these innovations more understandable, applicable, and useful to leaders and managers such as yourself.

 6. Learn from Natural Systems

Business ecology is based on the organizing elegance of natural systems, the success secrets that have accumulated over 3.5 billion years of evolution. Business ecology, by emulating natural systems design, offers several answers: articulating your organization's core genetic code, i.e., core purpose and values, life-cycle thinking, resource exchange relationships, and models for organizational design that encourage innovation, resiliency, and adaptability. These models are fractal, they apply to all sizes and scales of business and organization and are relevant for both day-to-day management as well as long-term strategic planning.

 7. Put Community Back in Business

Business ecology puts "community" back in business. Community lies at the heart of a healthy, life-sustaining business or organization, and is the essence of our spiritual well-being. Business ecology redefines success and is a powerful agent for restoring community within companies and organizations and the communities of which they are a part. Its mindware for sustainable enterprise offers models based on natural systems that incorporate the need for community among successful, surviving organisms— and organizations. Community is all about connections—to ourselves and each other.

WEAVING PROVEN SUCCESS STRATEGIES WITH LEADING-EDGE THINKING

Business ecology weaves proven success strategies from business and management leaders with leading-edge thinking. Key concepts that give a context to business ecology—such as sustainable development, organizational learning, ecological economics, and values-based strategic planning—are shown in Figure 1.1 and introduced in this chapter. While much of the discussion here and throughout this book centers on businesses, business ecology also applies to nonprofit, government, and community organizations of all types and sizes, and scales of operation.

Sustainable Development

Sustainable development, which came into the global and American lexicons in the late 1980s, is the successful integration of economic, social, and environmental goals to ensure the quality of life today and in the future. It is a long-term, values-based, interactive process that involves all stakeholders. Table 1.1 illustrates how various stakeholders might view sustainable development.

Figure 1.1   Weaving Proven Success Strategies with Leading-Edge Thinking

Table 1.1　Various Views of Sustainable Development

Bureaucrat	Not my job, not many resources are attached to this issue.
Engineer	Problem is too poorly defined. We can't compute results with precision. Come back to us when you have a *real* problem. Technology will have your answer.
Naturalist	The loss of species, and the rate of loss, should tell us something about what we are doing to Earth's life-support system.
Lawyer	What do you want it to be?
Economist	If there is a problem, get the price right and the market will resolve your problem.
Ecologist	Can a species in "bloom" be constrained by applying its own intelligence?
Systems Thinker	Incredibly intricate system with many degrees of freedom. A species should exercise caution when tinkering with complexity that it does not understand.
A Businessperson	Is there a profit? If not, don't bother me. I've got other business to tend to.
Native American	Man, earth, sky, and water are one. We all are bleeding.
Cosmologist	We know of no other planet that can sustain human life and the lives of species that we depend on. We must take care of Earth because it is all we have.
Artist	The answer is all around us, if we choose to take the time to listen and observe.
Cockroach	I've been through a lot before—I'll survive. These humans will disappear and the world will flourish once again. But I like all the nice snacks.

What do most people think about sustainable development? A nationwide Roper poll in 1996 found that most Americans (66%) believe that environmental protection, economic growth, and social equity are mutually supportive goals. The study, commissioned by S. C. Johnson & Son, more commonly known as SC Johnson Wax, discovered that most citizens see "win-win-win" outcomes as not only possible but preferable. This view differs markedly from the "either/or" mindset that has characterized mechanistic, industrial thinking and the political "jobs versus environment" debate.

What do businesses think about sustainable development? Business leader and author Paul Hawken sounded the call in 1993 with his book *The Ecology of Commerce* for businesses to use their power to create solutions for environmental problems around the world. In his book Hawken states:

> Business is on the verge of a transformation, a change brought on by social and biological forces that can no longer be ignored or put aside, a change so thorough and sweeping that in decades to come business will be unrecognizable when compared to the commercial institutions of today. We have the capacity and ability to create a remarkably different economy, one that can restore ecosystems and protect the environment while bringing forth innovation, prosperity, meaningful work and true security. The restorative economy unites ecology and commerce into one sustainable act of production and distribution that mimics and enhances natural processes.

Today, many companies in the U.S. and abroad see sustainable development as not only beneficial, but inevitable. Some, in fact, eagerly support sustainable development and are positioning for a strategic advantage. Robert B. Shapiro, chairman and CEO of Monsanto Company, made these remarks to Joan Magretta of *Harvard Business Review* (January/February 1997):

> Our nation's economic system evolved in an era of cheap energy and careless waste disposal, when limits seemed irrelevant. None of us today, whether we are managing a house or running a business, is living in a sustainable way. It's not a question of good guys and bad guys. There is no point in saying, If only those bad guys would go out of business, then the world would be fine. The whole system has to change; there's a huge opportunity for reinvention.
>
> We're entering a time of perhaps unprecedented discontinuity. Businesses grounded in the old model will become obsolete and die. At Monsanto, we're trying to invent some new businesses around the concept of environmental sustainability. We may not yet know exactly what those businesses will look like, but we're willing to place some bets because the world cannot avoid needing sustainability in the long run.

With all the public and private support for sustainable development, why is it not more widely applied? Carl Henn, Senior Vice President of

Royal Capital, is a business leader renowned for his vision and integrity. Based on his own experiences with sustainable development projects, he provided these valuable insights to the *Main Street Journal*, (published by the nonprofit Business Ecology Network).

> I have found that attempts to mobilize disparate organizations and movements to achieve greater collective momentum often fail despite similar basic goals. The usual well known reasons are that goals, strategies and priorities of some potential collaborators are not sufficiently similar, or, if the same, not much more can be gained working closely together than working separately. There are a host of other reasons that attempts to join forces fail as well.
>
> Why worry about it? Let each group do its own thing and collaborate actively with others only on appropriate, promising and well-defined win-win programs and projects of mutual concern. The whole sustainability thing is a learning process for everyone and there are many ways to learn. Different approaches tailored to relevant "local" circumstances tend to be more effective and probably attract a larger number of supporters overall.
>
> Good communications and networking are the key, as well as promoting a clearer understanding of the important relationships between different environmental issues and between these issues and other social problems of great concern (i.e., systems thinking). Many of these apparently different problems and issues have common origins. It's on these core cultural and political causes of a whole array of environmental and other problems that we should concentrate our collaborative energies.

Business ecology meets the challenges described by Hawken, Shapiro, and Henn. It is a new organizational paradigm that integrates profitability, stakeholder relations, and environmental goals—a win-win-win strategy for thriving in the ecological economy.

Organizational Learning

Leading thinkers such as Dee Hock, Arie de Geus, Peter Senge, Margaret Wheatley, and James Moore use natural systems as models for understanding and creating viable organizations. So too have James Collins and Jerry Porras in their work linking values with organizational success.

Dee Hock, founder and CEO emeritus of VISA International, is widely recognized for his original thinking about organizations. In the

1960s, Hock turned a fledgling credit industry into the hugely successful VISA global network, with an estimated trillion dollars in revenues in 1996. The secret to VISA's success? From its inception, Hock used biological metaphors to guide it.

Hock created a highly decentralized and highly collaborative organization modeled after natural systems. A dynamic tension was created, and still exists today, within the VISA system: competition among member financial institutions is balanced by the need to cooperate to make the system beneficial to all members. Today, Hock refers to VISA and similar organizational models as "chaordic" organizations that create a balance of chaos and order. More recently, he founded the Chaordic Alliance, a nonprofit foundation that supports his principles for organizational change.

According to Hock, "community" is the basis for any healthy organization. Our wandering from this basic natural model to a mechanistic industrial-age model is at the root of many of society's problems. Hock wrote in 1978:

> The mechanistic, command and control, industrial age corporation and the fixed manager for fixed duties, such godsends to the Industrial Age, are not only increasingly irrelevant, they have become a positive public menace.

Arie de Geus, renowned strategic planner and 38-year veteran with the Royal Dutch/Shell, believes organizational longevity is linked with biological principles, especially the instinct for self-preservation and improving one's condition. In *The Living Company*, de Geus writes:

> Like all organisms, the living company exists primarily for its own survival and improvement: to fulfill its potential and to become as great as it can be. It does not exist solely to provide customers with goods, or to return investment to shareholders, any more than you, the reader, exist solely for the sake of your job or your career. After all, you, too, are a living entity. You exist to survive and thrive; working at your job is a means to that end. Similarly, returning investment to shareholders and serving customers are means to a similar end for IBM, Royal Dutch/Shell, Exxon, Procter & Gamble, General Motors, and every other company.

De Geus highlights four characteristics of long-lived companies, based on his research with Royal Dutch/Shell:

1. *Sensitivity to the environment* represents a company's ability to learn and adapt.
2. *Cohesion and identity*, it is now clear, are aspects of a company's innate ability to build community and a persona for itself.
3. *Tolerance* and its corollary, *decentralization*, are both symptoms of a company's awareness of ecology: its ability to build constructive relationships with other entities, within and outside itself.
4. And I now think of *conservative financing* as one element in a very critical corporate attribute: the ability to govern its own growth and evolution effectively.

De Geus believes that living companies, organizations that emulate biological principles, have a strategic advantage in today's turbulent business environment. They are better able to respond to changes while maintaining a clear sense of identity. More traditional, mechanistic organizations, because of their rigid structure, lack of community, and poor learning skills, will simply not survive.

Peter Senge, author of *The Fifth Discipline*, agrees with Hock's and de Geus's observations. Senge, renowned for his work in systems thinking and organizational learning, helps organizations overcome "learning disabilities" that are developed and reinforced in our culture. Senge introduced five core learning areas to help people think and behave in new ways:

1. Systems thinking
2. Personal mastery
3. Mental models
4. Building shared values
5. Team learning

"The Fifth Discipline," the learning organization as Senge describes it, is the convergence of these five learning areas. Senge brought Hock to his MIT Center for Organizational Learning to help him discover how it can better serve its corporate members. Royal/Dutch Shell, Apple, Hanover Insurance, AT&T, and other members are developing their organizational learning skills to gain competitive advantage in the 1990s and beyond.

Margaret Wheatley, business consultant and associate professor of management at Brigham Young University, looks through the lenses of quantum physics, chaos theory, and molecular biology to show us how to manage organizations. In her book, *Leadership and the New Science*,

Wheatley postulates, for instance, that healthy organizations are made up of individuals who are self-referential. She states, "If management practice is ever to be simplified into one unifying principle, I believe it will be found in self-reference." She describes these self-referring individuals as those who find meaning in their work and life and are able to hold tightly to this "meaning" when under stress. They know how to yield—without breaking—during turbulent times and periods of organizational chaos.

Wheatley's discussion of chaos and our need to control it, as opposed to looking for its underlying sense of order, is based on her belief, which Dee Hock shares, that the "Old Science," Newtonian physics, has given us a flawed paradigm for viewing organizational management. This seventeenth-century perspective is linked to the creation of organizations that are hierarchical, rigid structures, object- rather than relationship-oriented, and incapable, in times of stress, of self-renewal and change.

By documenting ways of applying the "New Science" to organizational systems management, Wheatley builds a vision of organizations that are capable, like natural systems, of elegant solutions. They have minimum structure, yet maximum vitality, flexibility, and resilience. Chapter 2, "Giving Your Organization the Natural Edge," provides numerous examples of how applying such principles can give your organization a "natural edge" in the twenty-first century.

In *Built to Last*, James Collins and Jerry Porras provide an extensive analysis of what distinguishes truly exceptional, "visionary" companies from other companies that, although successful, are not quite in the same class in terms of resiliency, enduring excellence, and long-term shareholder performance. In fact, according to their study of eighteen pairs of visionary and comparison companies during the period from 1926 to 1990, one dollar invested would yield $415 in the general stock market, $955 for the comparison companies, and $6,356 for the "visionary" companies. Visionary companies delivered six times what the comparison companies produced, and over fifteen times what the general market produced.

James Collins, with previous experience with McKinsey & Company, Hewlett-Packard, and Stanford University, is an educator who runs a management learning center in Boulder, Colorado. Jerry Porras, who is the Lane Professor of Organizational Behavior and Change at Stanford University, previously worked for General Electric and Lockheed. Collins and Porras define visionary organizations as meeting the following criteria:

- Premier institution in its industry
- Widely admired by knowledgeable business people

- Made an indelible imprint on the world in which we live
- Had multiple generations of chief executives
- Been through multiple product (or service) life cycles
- Founded before 1950

Here they refute the myth that maximizing profit for the shareholder is always the top goal of "visionary" companies:

> Contrary to business school doctrine, "maximizing shareholder wealth" or "profit maximization" has not been the dominant driving force or primary objective through the history of the visionary companies. Visionary companies pursue a cluster of objectives, of which making money is only one—and not necessarily the primary one. Yes, they seek profits, but they're equally guided by a core ideology—core values and a sense of purpose beyond just making money. Yet, paradoxically, the visionary companies make more money than the more purely profit-driven comparison companies.

What makes these visionary companies so successful? Their enduring success and resiliency through multiple product life cycles and several generations of leaders are attributed to clearly defined core purpose and values being put into practice. Collins and Porras emphasize that these are visionary organizations that may or may not have visionary, charismatic leaders. It is clear from their study of visionary companies, such as 3M and Hewlett-Packard, that **core purpose and values, or what business ecology calls "social DNA," are linked to developing short- and long-term strategies, professional excellence in daily business activities, and adapting successfully to the dramatically changing and increasingly competitive business environment.**

A key to the success of visionary companies is that their behavior and decisions be aligned consistently with core purpose and values. According to Collins and Porras, a commitment to its core purpose and values—whatever they may be—is what distinguishes a visionary company from other companies in their industry. This means, while other things may change, decisions and actions should always hold true to core values. In many cases, this alignment is so strong that these organizations often exhibit cult-like characteristics, such as indoctrination, elitism, selectivity, and a fanatically held ideology.

Collins and Porras also dispel the myth that visionary companies necessarily use brilliant or sophisticated tactical or strategic planning, and that, in fact, their success mimics biological evolution:

Visionary companies make some of their best moves by experimentation, trial and error, opportunism, and quite literally—accident. What looks in retrospect like brilliant foresight and preplanning was often the result of "Let's just try a lot of stuff and keep what works." In this sense, visionary companies mimic the biological evolution of species. We found the concepts in Charles Darwin's *Origin of Species* to be more helpful for replicating the success of certain visionary companies than any textbook on corporate strategic planning.

They also point out that their findings on values-based organizational success are applicable to all human institutions—all types: small or large, nonprofits, government agencies—that seek sustained success. It is clear that core purpose and values are, in fact, social DNA for visionary companies, acting as genetic blueprints for organizational development, decisions, and actions.

James Moore, author of *The Death of Competition*, suggests that the biological model is already being applied extensively within the business community. He believes companies like Intel, Microsoft, and IBM are forming strategic alliances to gain advantage in their highly competitive business environment. According to Moore and others, the future belongs to companies that embrace interconnectedness and interdependence. According to Moore, they are the leaders and developers of "business ecosystems."

Esther Dyson, president of Dyson-EDventure Holdings, Inc. was quoted in the book as saying this about Moore's *Death of Competition*:

> Moore catches the fundamental shift in business thinking—and behavior—today: the economy is not a mechanism, businesses are not machines. They are coevolving, unpredictable organisms within a constantly shifting business ecosystem that no one controls. . . . Managers of companies both great and small must figure out how to evolve in this changing environment—to compete with what the competition is becoming, not with what it is now.

Business Ecosystems and Industrial Ecology

A business ecosystem is a dynamic community of companies and other "organisms" that echoes natural systems. Cyclical and mutually beneficial relationships develop within and among firms, institutions, and communities to share vital resource and commercial flows such as money, energy, materials, information, water, food, and people. Surplus flow from one entity is a valuable resource for another.

The Kalundborg, Denmark, business community provides an excellent example of an industrial ecosystem. Residual heat and steam from a power plant is used for heating nearby greenhouses, aquaculture facilities, a petroleum refinery, residential homes, and a pharmaceutical plant. Surplus gas from the refinery is purchased by the power plant—offsetting the need to burn up to 30,000 tons of coal per year. Sludge from the pharmaceutical plant, with lime added to neutralize pathogens, is spread for free on local farms. Recycling this sludge benefits the farmers by reducing fertilizer costs while providing an inexpensive, but effective solution for the pharmaceutical plant. Gypsum, a byproduct of the pollution controls (scrubbers) on the power plant, is provided to a Sheetrock™ manufacturer located on adjacent land. Waste Sheetrock™ from construction is also fed into this facility.

Business ecosystems such as Kalundborg, Denmark, can yield significant results for a company, community, and region. What are the potential benefits? They include increased revenues, a revitalized economy with new jobs, new products and markets, lower operating costs and liabilities, improved resource-use efficiency, healthy stakeholder relationships, pollution reduction and prevention, enhanced natural and cultural resources, and a higher quality of life. Chapter 5 describes business ecosystems from around the world. It also describes what we can learn from agro-ecosystems, and highlights challenges and opportunities related to business ecosystem development.

It is useful to compare business ecology with industrial ecology, to gain insights into the unique offerings of business ecology, as well as how it differs from industrial ecology. The intent of the authors, in making this comparison, is to anticipate questions of this nature, and to advance the knowledge and practice of sustainable enterprise. The following discussion is provided in that spirit, and is not meant to detract from the considerable knowledge, tools, and expertise developed by the industrial ecology community.

First, here is a definition of industrial ecology from the *Journal of Industrial Ecology*, a multidisciplinary, peer-reviewed quarterly:

> Industrial ecology is a rapidly growing field that systematically examines local, regional, and global uses and flows of materials and energy in products, processes, industrial sectors, and economies. It focuses on the potential role of industry in reducing environmental burdens throughout the product life cycle from the extraction of raw materials, to the production of goods, to the use of those goods and to the management of the resulting wastes. The field encompasses:

- Material and energy flows studies ("industrial metabolism")
- Dematerialization and decarbonization
- Life-cycle planning, design, and assessment
- Design for the environment
- Extended producer responsibility ("product stewardship")
- Eco-industrial parks ("industrial symbiosis")
- Product-oriented environmental policy
- Eco-efficiency

The above definition is a synthesis of concepts, research, and perspectives from the industrial ecology community. Based on this definition, **business ecology is considerably broader and deeper than the field of industrial ecology** in a number of ways.

Business ecology takes a holistic view of vital flows and relationships that sustain economic activity at multiple levels. In addition to energy and materials, business ecology recognizes other vital flows such as information and ideas, money, products and services, people and other organisms, air, water, and food (see Chapter 4). These cyclical flows contribute to value creation at multiple levels including products and services, individuals, processes, firms, communities, and economies. Business ecology also considers a broader web of stakeholders—those people that affect, and are affected by, an economic activity or entity (see Chapter 6). For example, a company's stakeholders can include customers, shareholders, creditors, suppliers, manufacturers, distributors, utilities, retailers, citizen and community groups, various levels of government, contractors, accountants, lawyers, competitors, and future generations. These vital flows and stakeholder interactions comprise the dynamic value-creation process within a firm, community, or business ecosystem. This process, as discussed in Chapter 2, includes creating a formative environment for ideas and innovations; planning, research, and development; purchasing, acquisition, and outsourcing; design and testing; production and processing; and marketing, sales, and distribution.

Business ecology is, at its core, values-based. Chapter 2 describes how values, as "social DNA," control organizational, technological, and economic development and determine who and what is valued. Recreating our organizations and economic systems requires a shifting of values, perspective, and behavior (see Table 2.1). Braden Allenby, renowned industrial ecology expert and AT&T's vice president for environment, health, and safety, highlights the need to address values and the limits of the current industrial ecology (IE) framework in *Business and the Environment* (October 1997):

. . . IE supports the scientific/engineering dimension of sustainability, but what kind of sustainability do we want? The whole dialogue on values hasn't occurred yet. It's difficult on the science side to go from point A to point B if B isn't defined.

Many leaders and thinkers, such as Dee Hock, Hazel Henderson, and Margaret Wheatley, contend that industrialization is at the heart of most of society's problems. Hazel Henderson notes succinctly in *Politics of the Solar Age*:

> The industrial model in its mass-consumption, global-advertising stage of titillating rural populations with visions of city lights, cars, flashy clothes, cigarettes, booze, Coke, and rock music now creates mass migrations throughout the planet in search of money, jobs, and status symbols. At the same time, traditional values appear old-fashioned, boring, and backward, described by Marx, who despised peasant culture, as the "idiocy of rural life." But after tempting every rural community on the planet that can be reached by mass media and transistor radios with the consumer model of the industrialized "good life," industrialism is caught in a cruel hoax: it cannot deliver. It can deliver for some of the population at the expense of others—and as we are now seeing, it can deliver only some of the time (while there are cheap, abundant environmental resources to be used up). But it cannot work in the long run.

Business ecology recognizes that the industrial system is simply one type of economic system, and that it is not sustainable. An industrial ecosystem is simply one form of a business ecosystem. Further, a "business" is more encompassing and longer-lived system than an "industrial" one. Restaurants, farms, and hotels, for example, can and do participate in business ecosystems and existed prior to the industrial revolution.

Making modifications to the industrial system while an ecological economy is emerging is equivalent to making design adjustments to the typewriter while the computer age is dawning. There is only so much one can do with a typewriter. It continues to serve a purpose, although greatly diminished, from its former, wider utility. But it is not a computer. Similarly, the emerging ecological economy is fundamentally different from the industrial economy. In the ecological economy, "ecological thinking" will dominate as "mechanistic thinking" dominated the industrial age. We will still need and use machines, but these machines will increasingly resemble living systems. For example, "smart" information technologies are already resembling sensory and neural systems of organisms. Industrial ecology may represent a "last gasp," an elegant innovation to a system that is being replaced. Industrial ecology is a step in the right direction, but at this stage we now need to take a quantum leap forward.

Creating a sustainable future requires a fundamental rethinking of human systems, including organizations, technologies, and economic systems. The industrial-age economic system, as so many of today's visionaries have observed, undermines the social, economic, and environmental systems upon which we depend. A new values system, which integrates business, community, and ecology, is already evolving in many sectors of society. In business and organizational management, more and more leaders are recognizing the flaws of the industrial-age economic system and its accompanying organizational models which do not take into account the fact that the Earth's resources and life sustaining capacities are limited. Robert Shapiro of Monsanto, in an interview with *Harvard Business Review*, describes how sustainable development is a major discontinuity that will drive business decisions:

> Sustainable development is one of those discontinuities. Far from being a soft issue grounded in emotion or ethics, sustainable development involves cold, rational business logic.
>
> This discontinuity is occurring because we are encountering physical limits. You can see it coming arithmetically. Sustainability involves the laws of nature—physics, chemistry, and biology—and the recognition that the world is a closed system. What we thought was boundless has limits, and we're beginning to hit them. That's going to change a lot of today's fundamental economics, it's going to change prices, and it's going to change what is socially acceptable.

Tachi Kiuchi, managing director of Mitsubishi Electric Corporation and former chairman and CEO of Mitsubishi Electric America, and Bill Shireman, president of Global Futures Foundation, invited leaders and innovators to join their "Future 500" network in 1997:

> Within a generation, Mitsubishi Electric and most of the world's major corporations as we know them will be extinct. Some may be out of business. But most will break out of our restrictive twentieth-century skins, and take on new forms which may be radically more productive and profitable.
>
> The change is not a matter of choice. To survive and thrive in the new economy, the global marketplace demands that our companies be structured less like machines, and more like dynamic living systems,

able to adapt to change, seize opportunities, and more fully capture the productivity breakthroughs enabled by information technologies.

Yesterday's dominant industries profited by transforming raw materials and fuels into products and services. To do so, we organized our companies like they were machines. The bigger the machines, the faster they could turn more resources into products. Standardization, uniformity, and independence became important profit principles, accelerating the speed with which the machines could operate. Strict top-down hierarchies dictated that all the machine parts operate according to specifications. Individuality had to be sacrificed for the smooth operation of the whole.

But the factors that maximized profits in the past—huge manufacturing plants, top-down hierarchies, conformist employees and customers, and uniform mass market products—now often strand capital in archaic systems that efficiently produce things people no longer really want.

Today, a new generation of CEOs and entrepreneurs is stimulating profits another way, discarding the machine model and structuring their companies like they were living systems. The living systems model enables them to change more quickly, seize new opportunities as they arise, and take greater advantage of the productivity breakthroughs enabled by information technologies.

Business ecology is based on the elegant structure and principles of natural systems. It recognizes that to develop healthy business ecosystems, leaders and their organizations must see themselves, and their environment, through an "ecological lens." This is how business ecology expands an organization or business's perspective from "cash flow" to "life flows" and from "accountability to shareholders" to "accountability to stakeholders."

Business ecology is both art and science. It incorporates the ability to see patterns, systemic relationships, different perspectives, and it builds mindware, organic models and designs for organizations based on close, scientific observation of natural systems. Art and science, working together, can improve our lives. Their artificial separation, and our separation from nature, stems from the fragmented, mechanistic thinking of the industrial age. Business ecology synthesizes these perspectives into one lens.

In summary, business ecology is broader and deeper than the field of industrial ecology. It is a comprehensive, values-based organizing framework that links profitability, stakeholder relations, life-cycle thinking, and environmental performance. As a catalyst for achieving sustainable development, business ecology is applicable and robust at different scales and types of human systems, including organizations, technologies, and economic systems.

Ecological Economics and Accounting

The economy is the environment for the business organism. It determines the rules of the game—how, what, and where products, services, and resources are bought, sold and exchanged, and who succeeds and who fails. Government taxes, laws, regulations, and subsidies, business practices and policies, prices and markets, cultural and political forces, availability of capital and resources, and the needs and desires of customers are examples of key drivers affecting the economy.

Ecological economics, sometimes referred to as "bionomics" or "ecolonomics," is a policy field that explores how the economy can be made consistent with natural laws, rules, and limitations. It involves policies such as tax reform, product stewardship, and incentives that encourage economic systems to behave more like ecological systems. Ecological economics seeks to "internalize" the "externalities" of the industrial economic system. An ecological economy is the environment of sustainable enterprise. While there are many opportunities for integrating economic, social, and environmental goals, the current economic system must be realigned to truly achieve sustainable development. The President's Council on Sustainable Development, for instance, a consortium of business, government, environmental, and community leaders, recommended the following in 1995:

Subsidy Reform
Redesign or eliminate federal subsidies that fail to incorporate the economic value of natural, environmental and social resources into the marketplace and into government policies.

Revenue-Neutral Tax Shift
Shift taxes away from activities that promote economic progress—such as work, savings, and investment—toward activities that lead to excessive environmental damage.

These are two examples of economic policies that could create a healthier environment for sustainable enterprise. Such incentives catalyze profitability, stakeholder participation, and environmental performance, helping business, society, and the environment "win" simultaneously.

Business organisms can and do affect their environment in ways analogous to how organisms shape their surrounding ecology. Humans are not alone as creatures who modify nature for their benefit. For example, beavers build dams and lodges by using trees, mud, and other resources. In similar ways, the economy must be shaped in ways that nurture sustainable enterprise. In this environment of sustainable development, profits, stakeholder participation, and environmental performance are mutually supportive and inextricably linked. Natural resources are valued for the services they provide and ecological restoration makes business sense.

Ecological accounting applies ecological economics to the scale of an individual firm. An ecological perspective of business can also transform the way we think about accounting and business management. Within our abstract economic system, it is easy to forget that cash flow is really a surrogate for other flows that sustain life, including the vitality of businesses. These life-sustaining flows within and among firms, households, institutions, communities, and the natural environment include materials, energy, food, water, air, people, and information. Chapter 4 provides examples of how looking at these and other flows through the lens of ecological accounting can help improve your organization's performance and work environment.

Within our industrialized society, it is easy to ignore or take these vital flows for granted. We have become isolated, or so we think, from the natural economy. It is as if we have forgotten where things come from or where they go. Business ecology provides a systemic framework for rediscovering these vital relationships. It helps us reconnect to the life-sustaining flows that support us, our businesses, our communities, and the web of life.

More comprehensive accounting of these flows provides a better picture of business health and supports the development of business ecosystems. Designing and creating business ecosystems, such as the Kalundborg, Denmark, example, can lower operating costs and liabilities, create market synergies, incubate new businesses, prevent or reduce pollution, and help revitalize communities and regions.

Good examples of comprehensive accounting are energy audits and the broader-based ecological audits, often called eco-audits. Utilities such as Pacific Gas & Electric Company, Inc. (PG & E) and private consulting groups such as Energy Conservation Management, Inc. of Baltimore, Maryland, are making profits by discovering waste energy and retrofitting products, production and distribution processes, facilities, and communities to improve efficiency and ultimately cash flow. Eco-audits, conducted by groups such as HVS Eco Services of Mineola, New York, simply extend the systems-based accounting to other flows such as water, materials, and information. Many consider this demand-side, ecological accounting and management a natural extension of total quality management. New opportunities for economic efficiency and pollution prevention are realized by adopting a "whole systems" perspective rather than the more traditional accounting methods that focus exclusively on cash flow and the bottom line.

Values-Based Strategic Planning and Accountability to Stakeholders

Strategic planning is essential to the business organism. It helps define the business environment and the many relationships and stakeholders affecting your company's health and performance. It includes: anticipating long- and short-term change; creating an organizational vision and goals; and identifying obstacles and opportunities related to achieving success. Strategic planning allows you to embrace uncertainty and, when done well, navigate your organization into position for competitive advantage. When linked with values—fixed points of reference much like the stars that aid navigators and sailors—strategic planning builds an organizational culture that inspires and motivates.

More and more organizations are discovering the benefits of values-based strategic planning and a broader sense of accountability to stakeholders. Here's an example from the Monsanto Company, based in St. Louis, Missouri, that illustrates dramatically what strategic planning can accomplish.

Monsanto scientists have come up with a number of elegant solutions directed at sustainable enterprise. One is the NewLeaf Potato™. It is bioengineered to defend itself against the destructive Colorado potato beetle. NewLeaf Potato Plus™, another new product, helps the potato fight a common scourge, the leaf virus. These two product innovations are linked to values of sustainability, profitability, and efficiency. They will eliminate the manufacture, transportation, distribution, and aerial spraying of millions of pounds of chemicals and residues. Monsanto scientists estimate that only five percent of the applied pesticides actually reach the target pests. The rest is dispersed into the environment. The use of significant amounts of energy, inert materials, insecticides, packaging, and containers will be avoided with the widespread use of these new products. Waste by-products and pollution are also eliminated.

Strategic planning requires us to ask probing questions about our work environment, helping us see obstacles and opportunities related to achieving our organizational vision. Understanding organizational values is a frequently neglected aspect of strategic planning. Why? Because they are not as easily quantified or measured as, say, bottom-line profits, many businesses probably lump values conceptually with personal development and community outreach. A discussion of values, however, is not peripheral to business. It is, in fact, central to organizational success. Examining values can uncover inconsistencies between organizational goals and the direction an organization is taking. It can also reveal the hidden, unspoken rules for success and failure that often run counter to a true sense of community, personal growth, and creativity. For example, a high technology company may define itself as innovative and cutting edge. However, a values discussion may show that politically savvy employees, rather than those who have original, cutting-edge and marketable ideas, are actually the ones who are promoted and recognized. The innovative, talented people may be leaving in search of better opportunities.

As described in Chapter 2, values are the natural blueprints, the genetic codes, of organizations. Self-referring organizations, as Wheatley and others suggest, have a true sense of identity. In Chapters 6 and 7, a systems-based tool is introduced as a self-referential map for values-based individuals and organizations. This tool enables an individual or organization to track progress toward identified values, and to evaluate or anticipate the consequences of a particular decision or action.

Business leaders such as Bryan Thomlison of Thomlison Strategic Alliances, Inc., Sue Hall of Strategic Environmental Associates, and Ralph Estes of Stakeholder Alliance, have demonstrated that moving from shareholder-driven accountability to a broader community of stakeholders can improve an organization's overall performance. In Chapter 6, their stories and others demonstrate how healthy stakeholder relations build healthy communities and profits.

WHY BUSINESS ECOLOGY MAKES SENSE

To understand why business ecology is, in fact, a logical evolution for businesses and organizations, it is useful to look briefly at the changing relationships among business, society, and the environment. Figure 1.2 provides four simple models to illustrate how these relationships have shifted dramatically, and why business ecology is a natural organizing framework for sustainable enterprise in the twenty-first century. In the first model, resources are mechanistically "flushed through," reaping disproportionate profits for industry and damage to the environment and communities. In the second, regulations are imposed to help bring balance to these inequities. In the third, closed-loop thinking, i.e. reuse and recycle, resource efficiency, and pollution prevention begin to emerge; this saves money for business and society and reduces the need for costly regulations. Finally, in the fourth model, business ecology provides a fundamental shift from mechanistic to ecological thinking, where viable business organisms are sustained by life flows, and profits, stakeholder relations, and environmental performance work together to create life-sustaining enterprise.

When examining businesses and organizations at a macro level, it is easy to see the fundamentals. Much like natural organisms, businesses "make a living" by transforming energy, material, and other life-sustaining flows into commercially viable products and services. In many businesses, "cash flow" is seen as the life-sustaining force and the machinery of production as the metabolic system. This profit-for-the-shareholder perspective affects how business is perceived, how success is defined, and how employees, natural resources, and other "factors of production" are perceived and managed. To break out of this organizational model, we must look at business through a new, ecological lens. A detailed look at the business climate and public policy that created the "flush through," "polluter pays," and "pollution prevention" models presented in Figure 1.2 is included in the Appendix. What is essential is that, in the business ecology scenario (Figure 1.2d), government's role changes dramatically from "enforcer" to "catalyst," and industry changes from "polluter" to "eco-efficient enterprise."

The concept of efficiency is something that industry understands very well. By looking at industrial flows with a new ecological lens, industry leaders are seeing their resource and energy inefficiencies. They have found that they can recycle both material and energy wastes back into the means of production, as shown by the closed loops of Figure 1.2c. As importantly, companies have found that their waste is another company's feedstock. Some of the waste stream, for instance, can be sold to businesses that use low-grade energy or material resources. Eventually, waste-exchange networks have emerged, representing a step toward business ecology. On the input side, these progressive businesses have streamlined and combined

Figure 1.2   (Opposite) **Four Organizational Models: Changing Relationships Among Business, Society, and Environment**

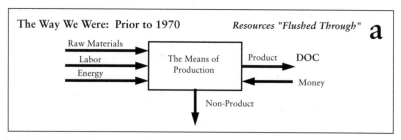

The Way We Were: Prior to 1970   *Resources "Flushed Through"* **a**

Raw Materials → The Means of Production → Product → DOC
Labor →
Energy →
← Money
↓ Non-Product

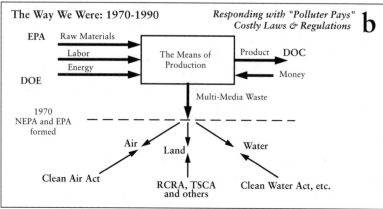

The Way We Were: 1970-1990   *Responding with "Polluter Pays" Costly Laws & Regulations* **b**

EPA   Raw Materials → The Means of Production → Product → DOC
Labor →
Energy →
DOE   ← Money
↓ Multi-Media Waste

1970 NEPA and EPA formed

Air   Land   Water

Clean Air Act   RCRA, TSCA and others   Clean Water Act, etc.

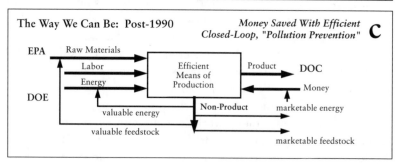

The Way We Can Be: Post-1990   *Money Saved With Efficient Closed-Loop, "Pollution Prevention"* **c**

EPA   Raw Materials → Efficient Means of Production → Product → DOC
Labor →
Energy →
DOE   ← Money
valuable energy   Non-Product → marketable energy
valuable feedstock   → marketable feedstock

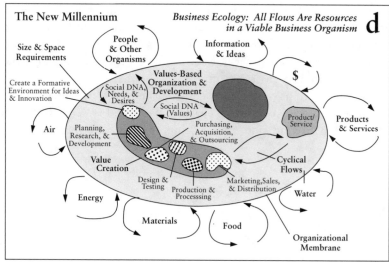

The New Millennium   *Business Ecology: All Flows Are Resources in a Viable Business Organism* **d**

Size & Space Requirements
People & Other Organisms
Information & Ideas
$

Create a Formative Environment for Ideas & Innovation
Social DNA, Needs, & Desires
Values-Based Organization & Development
Social DNA (Values)

Air
Planning, Research, & Development
Purchasing, Acquisition, & Outsourcing
Product/Service
Products & Services

Value Creation
Cyclical Flows

Design & Testing
Production & Processing
Marketing, Sales, & Distribution
Water

Energy
Materials
Food
Organizational Membrane

some of their separate business divisions. Combining, for instance, some of the functions of engineering, sales, and marketing with environment, accounting, and management, provides a new and simpler internal lens through which the company's means of production can be reassessed. Many companies, such as Dow Chemical, are finding that the more they look for pollution prevention opportunities, the more opportunities for profit they are finding—especially since most pollution prevention actions have payback periods of less than two years.

In this decade, closed-loop thinking has provided a new lens for seeing opportunities and a new channel of communication between industry, government, citizens, and the environmental community. Pollution prevention advocates and businesses have found that they have a common language— efficiency. By preventing pollution, companies can profitably reduce energy consumption, resource use, and costs for operation and maintenance. Most importantly, pollution prevention has pointed the way to more comprehensive, closed-loop solutions. Companies that embrace the pollution prevention ethic are finding that the rewards are positive and profitable; they largely can avoid the costly regulatory structure and its negative penalties by preventing pollution in the first place. Like business ecology itself, the closed-loop thinking of pollution prevention is a new way of seeing solutions that enhance sustainable business practices. It embraces the belief that the company's own talented people, properly motivated to creatively find solutions, are fully capable of solving environmental problems profitably.

Those who have fully embraced the pollution prevention ethic will easily see how business ecology works closely with closed-loop pollution prevention thinking with an even more holistic perspective. While pollution prevention focuses primarily on material and energy flows, business ecology considers all wealth-creating flows within a company. These wealth-creating flows include, for instance, the physical and mental energies of the employees, their sense of well-being, and flows of intellectual property. Figure 1.2d shows these and other flows in a simple way, as sustaining a viable "organism" within its environment.

As other sections of this book show, business ecology integrates both common sense and the organizing elegance of natural systems. It is a natural strategy for thriving, not just surviving, in the next economy.

2

Giving Your Organization the Natural Edge

The Earth has music for those who listen.

William Shakespeare

Any design, including the design of organizations, starts with a concept or idea that is shaped by a vision of reality and of a desired outcome. Working with the business ecology model for organizations, this chapter:

1. Looks more deeply at natural systems to understand the design environment from which business ecology has emerged, with a focus on economic systems and organizations;
2. Examines how life-cycle thinking can transform the way you do business and improve your organization's ecological efficiency and viability;
3. Discusses the life-cycle of value creation, including how ideas develop into viable products and services and what contributes to value creation in a sustainable enterprise;
4. Shows how to use natural models to grow innovation within your organization.

A DEEPER LOOK AT DESIGN, ECONOMIC SYSTEMS, AND ORGANIZATIONS

Design is an expression of intent. Intent is formed by needs and desires. These, in turn, are shaped by our value systems. Values are the "social

DNA" for developing human systems, whether they be economic, techno-logical, or organizational.

In business, the term design is often applied to products, technologies, buildings, and software. Design also applies to economic systems. As men-tioned previously, the current industrial economic system evolved in an era of perceived limitless resources and growth. Throughout human history, diverse cultures and economic systems have evolved to meet the changing needs and desires of the individuals that comprise those systems. Many of these systems, including hunting and gathering, farming, feudal land use, colonial expansion, and the modern industrial system, are still in use today.

What are the values of each of these systems? How do these values influence the design of human organizations and enterprise? Here are some examples that show how values shape organizational design. These illus-trations raise an intriguing, paradoxical question: How much and how little have we and our systems really changed?

Just as hereditary characteristics filter down through the centuries, the values of hunting and gathering and farming cultures are part of the genetic code for today's organizations and enterprises. Even though the in-dustrial system is dominant in Western culture, we continue to be shaped by the social DNA from these simpler economic systems. In fact, values, as social DNA, enter into business, technological, and economic decisions all the time.

Since the beginning of human civilization, societies have developed different adaptive strategies for surviving. Many, such as traditional Eski-mos and Australian Aborigines, have lived harmoniously with nature. They continue to have a keen awareness of their environment, the cycles of life, and the organisms that sustain them. However, there have been and con-tinue to be warrior-like cultures that have skillfully and forcibly exploited and conquered nature and other peoples to their advantage. The Assyrians, the Vikings, and the Romans are a few well-known examples. If predatory cultures only took what they needed, there would be a balance and a chance for the environment and communities to rebound. However, once a culture, such as those mentioned and even our own, seizes more than it needs, a belief system is often created to sanction its excessive behavior.

Underlying values are at work here, values that are deeply woven into such a culture's worldview: nature and other people exist to be exploited; the Earth has unlimited resources; and what is taken will "somehow" be replaced. This is akin to the way a young baby or child views the world—believing that "everyone exists to take care of me"—until he or she ma-tures. In our world of sharply rising human populations, limited and diminishing natural resources, acute competition for those resources, and stressed ecosystems, these culturally self-centered values can be dangerous, even life-threatening.

On the positive side, many traditional cultures, such as hunters and gatherers, cooperate and specialize in several ways. These include male hunting parties, female foraging and childrearing groups, and elderly storytellers and shamans, the respected keepers of wisdom, medicine, religion, and tradition. In these cultures, the lines drawn by kinship and taboos are strong organizing forces. Values of cooperation, cohesion, tradition, and specialization within a tribe are balanced by competition for status within the hierarchy, competition and even brutality among different tribes competing for the same resources, and a general working knowledge of nature, including the ability to adapt and survive in an ever-changing environment. Storytelling and various rituals create a thread that connects these values from one generation to the next in much the same way that the Bible, for instance, has kept alive the history and values of some of the world's modern-day religions.

Agriculture evolved around several of the cooperative values held by early hunters and gatherers, and introduced new ones as well. Values, such as connection to place, stewardship of the land and living resources, and understanding the natural cycles for planting and harvesting, emerged along with more complex, less nomadic forms of community. This rooted lifestyle created conditions for the advancement of trade, technology, language, art, villages, cities, and other dimensions of what has today become our modern industrialized cultures and economic systems.

These economic systems are shaped by core cultural purposes and values, or social DNA, as well as by necessity and desire. What is valuable is a relative thing—molded by the worldview of the individuals who sustain the system. **In fact, two illusions of our times are (1) that our economy is value-neutral, and (2) that people and organizations have sufficient information to take full responsibility for their decisions.** For example, how many people, when they purchase gasoline, consider whether the price at the pump truly reflects its cost of production and distribution? Or think about whether purchasing one type of fuel over another reflects a values choice? Does the purchase of gasoline—and the price—reflect the cycle of environmental costs and military insurance that keeps oil coming from the Middle East? Hazel Henderson writes in *Politics of the Solar Age*:

> We are coming to realize that, ultimately, all social control systems operate at the level of language and symbol systems, encoding in various cultures their values (i.e., what is valuable and who is valuable).

Then the valued people and activities are drawn inside the magic circle of monetization, while those devalued are left out. Thus, any economic system can appear successful, depending on where such boundaries are drawn. Thus, crucial relationships exist (often denied and rationalized) between culture and ethics and all economic/technical systems. *Value systems and ethics, far from being peripheral, are the dominant, driving variables in all economic and technological systems.* It is in this sense that Marxists assert, correctly, that all knowledge is political. Similarly, all science is value-based.

Thus, the task facing industrial societies as they enter the 1980s and 1990s and their coming "trial by entropy" will be to face up to the *unsustainability of their value systems*—rather than view their "problems" *as deficiencies of nature.* This kind of "gestalt switch" out of our infantile, anthropocentric preoccupations is now the prerequisite for the survival of our species.

Stated simply, it is our turn in the cycle of world history to "grow up" and start responsibly taking care of the resources, living systems, communities, and people that have been taking care of us. But can individuals and their larger communities and organizations create the shared values necessary to motivate such supportive, responsible actions? It is difficult when values, such as resourcefulness, nonviolence, thrift, and honesty, conflict with the dominant economic system and culture at large.

Organizations, like organisms, are shaped by both their internal social DNA and the social and economic environments in which they live. Values, as social DNA, determine organizational design because they are one of the key motivators that energize the individuals that make up an organization. With each decision a value-driven employee makes, his or her organization then resonates with a tangible identity. In the best case, employees and organizations share the same values, the same social DNA. In this case, an organization works together as a vibrant, self-organizing system with clear direction and purpose.

As pointed out by Dee Hock, Margaret Wheatley, and others (see Chapter 1), the mechanistic organizational model that we have inherited has roots in the 1700s and earlier periods. Why has organizational evolution lagged behind, while other dimensions of society, such as science, technology, and communication, have progressed considerably? Perhaps because organizational design demands a systemic, holistic approach, which is not one of our cultural strengths. The other fields are more specialized and thus more readily developed by our Western culture, which is predisposed to linear ways of developing information.

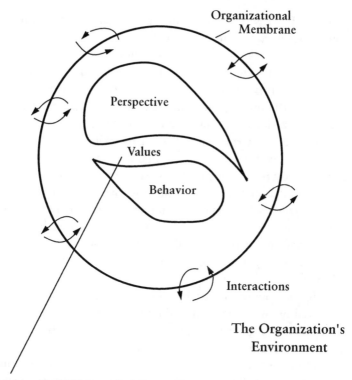

Organizational values, like DNA in a cell, define membrane structure, determining what is filtered in and out. These values are affected by their environment in ways that mimic living systems. No living organism can be sustained in the long term in isolation from its environment.

Figure 2.1   An Organic Model for Organizations, Their Design, and Evolution

© 1998 Business Ecology Associates

How can we change? Organizational inertia, or an unwillingness to change, is tied to three powerful, mutually reinforcing forces—values, perspective, and behavior (Figure 2.1). Each of these affects, and is affected by, the other two. Values, as social DNA, shape the lens through which we see the world and are the basic blueprints for decisions, actions, and organizational development. *Perspective*, as the lens through which we see and interpret reality, in turn, affects how our values develop and how we behave. *Behavior* is how we act and relate within our environment. Environmental feedback regarding our behavior shapes our values and our perspective. **If we wish to recreate our organizations, we must change our values, perspective, and behavior.**

We often fear change because a new idea or approach challenges our perspective, or mental model, of how the world works. Embracing change often means leaving the "comfort zone" of our strongly-held perspective and reconstructing a new truth, a new perspective. **Yet those individuals and organizations that embrace change thrive in today's uncertain, turbulent business environment.** The fast-growing Xerox Business Services (XBS), described in Chapter 3, is a good example. XBS's change strategy—and its success—is shaped by its core purpose and values, its social DNA. These include an energetic, creative learning environment; shared vision; building trust, self-knowledge, and healthy relations; and seeing organizations as natural systems. By sharing their customers' habitats and tapping the creative energy of its 15,000 employees, XBS adapts quickly to its customers' changing needs, anticipates problems, and develops timely, effective solutions.

In the context of organizations, most of us have been educated and employed by institutions shaped by the mechanistic model. To breathe life back into our organizations, we must change and discard, when necessary, the mechanistic model to embrace an organic one.

How can we identify the steps for making such a profound organizational change? **Business ecology is *mindware* for transforming your organization into a sustainable enterprise.** Business ecology takes the first step of defining a common language that links the past with the future. Business ecology is an organizing framework for change. It spotlights those intangible elements of organizational design, such as values, perspective, behavior, and even culture, that are essential. Business ecology shows the key relationships that link these intangible design elements to concrete systemic results, including profitability, stakeholder relations, and environmental performance. It is not unlike discovering, for example, that the forces that make a traditional tribe thrive successfully in its environment are its deeply held kinship patterns, taboos, and beliefs—not the market price of or demand for their animal skins and jewelry. This is what mindware can do: it allows you to see simply and systemically how your organization really works.

Figure 2.1, an example of mindware, shows how external influences from the "outside world" or environment may affect organizational evolution. An organizational membrane is shown to define the boundary of the "organism." The thicker and less permeable the membrane, the greater the isolation from external forces. The danger, of course, is that the "organism" loses touch with the conditions outside of itself. It becomes a world upon itself, removed from life-sustaining flows that nurture creativity or the stakeholder needs that shape its services or products, and removed from a sense of responsibility to its community. It may create its own rules, regardless of what is known to be true outside. When the membrane gets thick and impervious, the organism becomes an isolated, self-contained system. Such systems are stuck in their internal paradigm and find it difficult to perceive or adapt to changes in their external environment. Just as "no man is an island," no organization should be isolated. Businesses and organizations need active networks of relationships and communication to keep growing and adapting, rather than stagnating or solidifying into a "denial" state.

This "denial" explains, in part, why, despite the warning signs, industrialized society continues to "consume" despite the obvious cost to our social, economic, and environmental systems. Figure 2.1 also adds weight to another explanation as to why organizational evolution has lagged behind while other aspects of society, such as science and technology, have progressed considerably: many organizations have developed protective membranes that isolate them from each other and their environments, locking them into self-reinforcing value systems that are poorly suited for today's challenges and opportunities. Their internal environments are shaped by the Newtonian mechanistic, industrial-age model, while their membranes filter only that information that reinforces this worldview. It is also obvious that an organism that continues to foul its own environment— while pretending to be isolated and self-sufficient—cannot be sustained in the long run. In fact, no living organism can be sustained in isolation from its environment.

Creating healthy, life-sustaining organizations means creating organizations that dynamically interact with the outside world and other organizations. Like a cell, organizations must exchange life-sustaining flows through a semipermeable membrane, selectively using what is needed and expelling or screening that which is not. **This selectivity can be vital to an organization's growth, adaptation, and long-term success. Organizational values, like DNA in cells, define the structure of the membrane, determining what passes through and what is blocked.**

Returning to Figure 2.1, organizational values affect and are affected by their environment in ways that mimic living systems. DNA, for example, has profoundly shaped the development of life and consequently the Earth itself. In turn, primordial Earth created conditions for life to begin and flourish, including the development of DNA. In the same sense, an organization, such as a business, industry, or community, is shaped by both its core purpose and values—its social DNA—*and* its environment.

The values of the mechanistic organizational model fit well with those of the industrial economy. In fact, this organizational model is the "organism" and the industrial economy is its "environment." They have coevolved and are mutually supportive of each other. This is yet another reason for their persistence. The following example shows how the industrial economy and its accompanying organizational model—the mechanistic model—often thrive at the expense of other organizational forms and economic systems.

The Brewing Business

Social DNA and the economic environment can shape an industry, including product choice, scale and type of production, and marketing, sales, and distribution. The following account, by Charlie Papazian, author of *The New Complete Joy of Home Brewing*, describes the history of the American brewing industry:

> What is American beer? Today's typical American beer is a lightly colored, light-bodied pilsner-lager beer, a style very different from the American beer of yesteryear. Through the years, it has been very much influenced by agriculture, climatic, economic, political and cultural factors.
>
> Before Prohibition, literally thousands of breweries existed, each supplying their respective regions with distinctive styles. There were, as well, millions of people homebrewing quality beer. The healthy diversity of beer styles must have been wonderful to experience. One imagines that there was a genuine sharing of kinship among brewers, whether they were homebrewers or professionals. It must have been that important feeling that went into the beer that made all the difference.
>
> Between January 1920 and December 1933, the United States suffered through Prohibition and the dark ages of beer. When it was over, only the larger breweries had survived by making malt products for the food industry. Low-budget operations combined with equipment left idle and in disrepair for over a decade eventually led to the demise of the smaller, local breweries.
>
> What was reborn was an industry of larger breweries. They were still somewhat anxious about the prevailing attitude toward alcohol. As incredible as this may seem, many of the richer styles of American

beer were not brewed, in an attempt by the breweries to market beer that would appeal to women.

Mass marketing began to rear its foaming head in search of the perfect beer that would appeal to most people. Never mind diversity. Never mind variety. Never mind the traditional ideals that American brewers had developed for more than 150 years. Never mind choice.

Then came World War II. A shortage of war material necessitated the scrapping of steel, some of which was idle brewery equipment. A shortage of food diminished beer production. The beer that was made had less malt in it. Many men were out fighting a war, and the beer drinkers back home were mainly women.

A lighter style of beer was thus beginning to gain popularity in the United States—and justifiably so. With the warm climate that we in the States enjoy for half a year, a lighter beer can be a refreshing experience. With the agricultural abundance of corn and rice here, these ingredients have found their way more and more into American beer, lightening the taste and body. If it's well brewed and you enjoy it, there is absolutely nothing wrong with this kind of beer. What we are missing now, sadly enough, is choice. The economics of mass marketing have indeed influenced what is offered.

Both this brewing industry example, and the aquaculture comparison in Chapter 3, illustrate "viral" or "parasitic" development when one type of social DNA overrides the existing values of another economic system. The brewing example shows how the Prohibition movement and other conditions created an environment for large industrial breweries to survive and later thrive, while most of the diverse, well-established, community-based breweries perished. Similarly, the comparison of aquaculture metabolisms in Chapter 3, illustrates how Western development policies created conditions for "modern, intensified techniques" to flourish at the expense of more "traditional" approaches.

Today, the rising popularity of microbreweries, brewpubs, and home brewing in the U.S. suggests a resurgence in the craft and diversity of brewing that was once a part of the American landscape. Europeans have managed to preserve their long-standing brewing traditions, despite the onslaught of mass-produced, mass-marketed beers. To them, it is more than producing great-tasting beer. It is about preserving values such as culture, heritage, diversity, and craftsmanship. In many European communities, brewpubs are also vibrant gathering places where families, co-workers, and neighbors interact and discuss important political and civic issues. In this sense, they represent a form of community, both abroad and in the U.S.

Table 2.1   Comparing Industrial and Ecological Economies

Shifting Values, Perspectives, and Behavior . . .

A. LESS OF THIS	B. MORE OF THIS
Linear Production	*Cyclical Systems*
Win-Lose	*Win-Win*
Extraction/Exploitation	*Symbiosis/Harmony*
Technology Controlling People/Nature	*Nature/People-Designed Technology*
Mechanistic	*Organic*
Mass Produced/Large Inventories	*Just in Time/Just in Place Quality*
Consumer	*Customer*
Market-Driven, Legalistic	*Values-Driven*
Remote	*Community/Context-Oriented*
Uniformity	*Celebrate Diversity*
Control/Manipulative	*Nurture Creativity/Feedback*
Economy of Scale	*Ecology of Scale*
Job	*Purposeful Work/Right Livelihood*
Compartmentalized	*Systemic*
Rigid/Resist Change	*Flexible/Adaptive*
Cash Flow	*Life Flows*
Profit as End	*Profit as Means*
Growth	*Development*
Amoral/Unethical	*Moral/Ethical*
Materialism/Products	*Sustainability/Service*
Competition/Predatory	*Balance of Competition/Cooperation*
Colonial/Dependency Creating	*Empowering/Nurturing*
Hierarchical	*Self-Organizing*
Fast-paced/Hurried	*Rhythms of Life/Spiritually Centered*
Accountability to Shareholders	*Accountability to Stakeholders*
Ego-Centered Leadership	*Service-Oriented Leadership*
Bottom-Line Accounting	*Integrated, Full-Cost Accounting*
Limitless Resources	*Limited Resources*
Near-Term Focus	*Future/Time Cognizant*

Comparing Industrial and Ecological Economies:
From Cash Flow to Life Flows

Table 2.1 compares the industrial economy with the emergent ecological economy. It is important to note that this table presents information in a black-and-white, linear fashion to illustrate a shift in thinking as we move toward the twenty-first century. In life, there are many gray areas and different ways of seeing. There are items listed in the left column, for example, that in many instances make sense for people and nature. Some degree of hierarchy, uniformity, and focus, for example, is often necessary to get things done. This table also conveys the importance of broader or different perspectives. For instance, cash flow is really a surrogate for energy, material, and other life flows.

It is useful to compare the industrial economy with the emerging ecological economy to highlight changes that have already taken place, trends that are just emerging, and the challenges that lie ahead. The industrial economy is consumption-driven, object-oriented, linear, and energy and material intensive. Demand for "things" is created through mass marketing. Enormous amounts of energy and materials flow through and support industrial production, with much of it wasted. Things are made, sold, and distributed to consumers and disposed of in a linear, one-way flow. Used things, and the associated mountains of waste, by-products, and packaging, are disposed of in landfills, incinerated, and otherwise "managed," with the costs generally borne by governments and taxpayers. Recycling, which has grown in recent years in the U.S. and abroad, represents a shift from linear to cyclical patterns and is an early signal of the developing ecological economy.

While the industrial-age economy is linear, extractive, and wasteful, the ecological economy is cyclical, symbiotic, and resource efficient. Business ecology provides a powerful lens for identifying and realizing opportunities for ecological efficiency and closing the loop. Residual flows, by-products, and used products become inputs into the next cycle of production, once sufficient use has been derived from the original products and services. Cyclical flow systems, as detailed in Chapter 3, support the exchange of life-sustaining flows, such as energy, information and ideas, product and services, air, people and other organisms, food, water, and materials.

While the industrial economy is object-oriented and profit- and consumption-driven, the ecological economy is relationship-oriented and values- and quality-driven. Building healthy relationships with stakeholders and vital flows is essential for organizational viability. As already noted in Collins and Porras's *Built to Last*, values-based organizations outperform strictly profit-driven ones and are better able to adapt to a changing, competitive business environment. Finally, value creation in the ecological economy seeks to meet the needs and desires of customers, with a focus on real quality-of-life improvements. Customers are active stakeholders who sustain an organization and inform the product and service life-cycle, so that just-in-time, just-in-place, just-for-the-customer outcomes are created.

In a closed-loop economy, customers pay for the value, or quality-of-life improvements, delivered by products and services that are produced through increasingly efficient use of life-sustaining flows. Products and services can go through several cycles of use before they are recycled back through the system. Once a product, for example, diminishes in value, it passes along a "food chain" of users until eventually it is disassembled or "decomposed" to begin the life cycle again. Here is an example from the computer industry.

Personal computers may cycle through several user groups, from high-end users who want state-of-the art equipment, to people who want lower-cost but quality computers, to small businesses and nonprofits who have limited budgets and minimal computing requirements. This cycling of ownership among different user-communities is driven by the advancement of computing technology and performance, pricing, and varying needs of customers. Eventually, the computers may be disassembled (or demanufactured) and recycled, with parts and materials being used in a new generation of computers or other uses. The next section examines how this kind of life-cycle thinking can help you position your organization favorably in the ecological economy.

THE POWER OF LIFE-CYCLE THINKING

Life-cycle thinking is a powerful tool for seeing opportunities at various scales—including product and service, organization, and economy—to improve ecological efficiency. Life-cycle thinking is gaining momentum in Europe and Asia and within global industries. *Product Stewardship Advisor* (May 1997) reports a flurry of "take-back" legislation to encourage product stewardship and life-cycle management:

> Several European nations are enacting a spate of new product takeback legislation that requires manufacturers to assume some responsibility

for what happens to their electronic products at end of life. There are also indications that something similar may happen in Asia.

Some leading companies are not waiting around for enactment of such legislation, but are aggressively moving toward closed-loop thinking and product stewardship for competitive advantage. Here is an excerpt from the same *Product Stewardship Advisor* report:

> Mitsubishi Electric America (MEA) and its Japanese parent company are acting on their convictions about product stewardship, and that is why MEA is considering operating a half-dozen private takeback centers across the U.S. within 10 years.
>
> A year ago, MEA combined its product quality and environmental management divisions into one entity, the environmental quality division. Other corporations that are adopting the ISO 9000 quality standards and ISO 14000 environmental management systems are performing similar combinations. For MEA, it represented a mindset that would come to incorporate stewardship concepts and life-cycle costs in product design.

Many businesses and governments are recognizing the value of shifting from linear to more cyclical behavior. Take-back legislation in Europe encourages businesses to be accountable for the life-cycle of the products that they produce. For example, once a car becomes inoperable, the car goes back to the manufacturer. A car manufacturer today becomes a "transportation provider" in the ecological economy. Economic incentives encourage the recycling of materials and energy within industrial systems. In the U.S. and elsewhere, recycling enterprise zones are sprouting up in response to constraints for municipal solid waste disposal and incentive programs that encourage new markets for recycled products. In fact, several of the business ecosystems profiled in Chapter 5 have evolved around recycling enterprises and financial incentives.

Merging Quality with Closed-Loop Thinking

The international community also recognizes that quality products, quality management and quality thinking go hand-in-hand. Starting in 1992, with the international ISO 9000 standard, the International Organization for Standardization expanded their role to include certifying that companies complying with ISO 9000 were using "best available practices" in the design and development of products. Contrary to popular belief, "ISO" is not an acronym, but refers to "equal measurement or quantity," as in "iso-therm" or "isobar." More recently, a new environmental standard, ISO 14000, seeks to provide a standard of excellence for companies and products that looks comprehensively at life-cycle uses of energy and materials. The ISO 9000 and 14000 standards are perhaps the most potentially far-reaching developments affecting the business community. In the context of ecological economics, they are self-imposed private sector policies that are reshaping the business environment, and like public sector policies, are likely to induce institutional behavioral shifts. The following excerpt from a meeting of the U.S. Interagency Environmental Technology Organization conveys these points:

> The ISO is a non-governmental organization established in 1947 to harmonize national and regional standards in order to facilitate trade. ISO standards are voluntary and are mainly technical in nature. The ISO 9000 (Quality Management Standards) is a process standard which provides a way for companies to establish a systematic and internationally recognized management system. ISO 14000 is a process standard in the area of Environmental Management Systems (EMS). ISO 14000 is a series of voluntary environmental standards in the areas of: Environmental Management Systems, Environmental Audits, Environmental Performance Evaluations, Environmental Labeling and Claims, and Environmental Life Cycle Assessment.
>
> ISO 14000 will potentially have wide-ranging impacts on how the government and industry manage, measure, improve and communicate the environmental aspects of their operations in a systematic way. The standards will influence the design, manufacture and marketing of products, the selection of raw materials, the types of environmental data that is gathered and how this data is communicated to governments and to the public.

Total quality management (TQM) introduced the notion of the value chain, where people within and outside an organization interact with each other and stakeholders. Depending on the situation, each individual can be both a supplier and a customer. With each transfer from supplier to customer, value is added to a product or service. For example, in the creation of a book, the authors supply a book proposal to the publisher, the initial customer. If accepted, the publisher counters with a contract, or a publication agreement that defines the terms of preparing a manuscript, the publication process, roles and responsibilities, intellectual property rights, schedule, and compensation, such as royalties and advances. The publisher supplies this publication agreement to the authors, who are now customers in the second link in the value chain. Once the publication agreement is approved and signed and the advance money disbursed, the authors create a manuscript that is eventually submitted to the publisher's editorial staff, marking the third link in the value chain. The manuscript, in turn, is edited and perhaps returned to the authors to make revisions. The value creation process continues with subsequent steps, such as proofreading, indexing, typesetting, graphic design, printing, and marketing and distribution, until it is bought by a customer in a bookstore or reading club. This process in its entirety is the life cycle of value creation, which is addressed more fully in a later section of this chapter.

TQM seeks continuous improvement along each link in the value chain, or life cycle, where the activities of each person are evaluated in terms of the value added to the process. Traditionally, quality and efficiency improvements have focused on products and services, profitability, and customer relations. In the context of business ecology, these quality and efficiency improvements extend to all life-sustaining flows and stakeholders. Time, space, and all life-sustaining flows are efficiently used to create optimal value for stakeholders. In the publication example described above, computers, energy, information and ideas, paper, inks, printers, mailing packages, pens, binding materials, and the creative energies, skills, and experiences of the authors, editors, graphic designers, distributors, and other professionals contribute to the cyclical, value-creation process. Examples of quality and efficiency improvements include electronic transfer of book-related files, research, communication and marketing via the Internet, using nontoxic, soy-based inks, using recycled paper and recycling used paper, and a fully conceptualized outline to guide the creation of the manuscript. In this last example, core principles—values and concepts—can act as the book's blueprint during the value creation process, thus focusing writing activities and minimizing waste of time, energy, materials, and other life-sustaining flows.

The value embodied in a book is dispersed among many stakeholders, including those profiled or interviewed in the book; individual readers; organizational customers, such as schools and libraries; authors; editors; the publishing company; the printing company; bookstores; and book clubs. The Internet and video and audio tapes are examples of by-products or derivative works associated with a book that create value by presenting the content in a different format, thus producing additional income streams. A community of stakeholders benefit from the original and subsequent value-creation activities.

Quality and efficiency improvements can create both positive and negative impacts on a stakeholder community. Systemic changes can be particularly unsettling, where stakeholders either adapt or perish. In the sound recording industry, for instance, the shift from vinyl records to compact discs stimulated change within a whole community of stakeholders, including the manufacturers of record-players, record stores, radio disc jockeys, and the record-buying public. Similarly, robotics in manufacturing industries have forced laborers, plant managers, and other stakeholders to retrain or face potential economic hardship. Business ecology provides a systemic lens for seeing and anticipating the consequences of quality and efficiency-related decisions on stakeholder relationships.

THE CHANGING BUSINESS ENVIRONMENT: ECONOMIC CYCLES

Economic cycles are broadly recognized by business and society, but there are as many theories over why they occur as there are economic analysts. The complex economic models used to explain or predict growth and contraction cycles are fundamentally rooted in the detached, industrial worldview. Consequently, these abstract models have little or no connection to natural laws and principles, generally ignore nonmonetized aspects of the economy, include only quantifiable terms, and generally mystify lay people not familiar with econometric jargon. Readers who wish to explore the theoretical basis of economics in the context of sustainable development are advised to read the works of Hazel Henderson, author of *Building a Win-Win World* and five other books. Henderson, who has served at the National Academy of Engineering, the National Science Foundation, the U.S. Office of Technology Assessment, and the University of California, Berkeley, provides an extensive analysis of the history and theory of economics, including the underlying assumptions of its models.

Thornton "Tip" Parker and Ted Lettes of Growth Cycle Design, Inc. in Bethesda, Maryland, have developed a practical, straightforward framework that demystifies why and how economic cycles occur. Their framework also fits well with the business ecology model and sustainable development. Here are highlights from their report, "America Needs Another Growth Cycle," which appeared in the Business Ecology Network's newsletter, *Main Street Journal* (Vol. 2).

Parker and Lettes explain economic growth cycles in terms of four aspects of the economy: dominant modes of transportation, dominant fuels, dominant materials, and dominant goals. In the first cycle, from the mid-1800s to early 1900s, U.S. industrial growth was defined by railroads; wood and coal; wood and iron; and westward expansion and the development of northern manufacturing and finance centers. Government was an active agent in this growth cycle, nourishing economic growth in many ways, including: subsidizing private construction of infrastructure, such as railroads and telegraph systems; creating lucrative opportunities for New England industries by placing tariffs on imported goods; and issuing land grants to colleges and private citizens.

A second cycle of U.S. industrial growth followed the first. Highways and air; petroleum and electricity; steel, concrete and aluminum; and suburban home building and expansion of southern and western manufacturing defined this second growth cycle, which began in the early 1900s. Again, the U.S. government played a major role. It funded or subsidized, for example, the advancement of aeronautical technologies, hydroelectric power, air navigation systems, computational design techniques, and highway and airport infrastructure, all of which provided an incentive for private sector commitments and spawned technological innovation.

Parker and Lettes also provide examples of how large companies benefited from these large-scale economic cycles and government investments by creating their own self-reinforcing growth cycles. Here they describe General Electric (GE):

> For over seventy years, General Electric had its own self-reinforcing growth cycle. One side of the company built equipment to generate and distribute electricity. The other side made everything it could think of to consume electricity. The growing consumption caused power companies to expand and buy new equipment. This lowered the cost of electricity which led to new applications. It is impossible to count all the products fostered by this cycle that contributed to science, education, health, transportation, manufacturing, communications, food, and everyday living. The cycle ran until the mid-1970s when stockholder demands and the oil embargoes forced GE to concentrate on increasing its stock price in the new era of energy conservation.
>
> GE and a few other large companies created their growth cycles by establishing strong positions in several industries that had to interact. Unlike conglomerates of unrelated parts, when one division of these companies grew, demands were created for the others. The companies were sustained by their growth cycles even when individual products failed. Countries have used growth cycles since the Industrial Revolution and some are doing it today.

Parker and Lettes discuss convincingly how the U.S. can lead a transition to a sustainable economy by creating a growth cycle for sustainable enterprise and technologies. Such a growth cycle would encourage value creation that is increasingly ecologically efficient; that is, it would use less time, space, and life-sustaining flows, such as energy, materials, and water in creating a higher quality of life for citizens in the U.S. and abroad. A sustainability-driven growth cycle would provide the necessary framework for long-term investments for individuals, companies, and other investors. Contrary to more speculative investments, substantial earning streams and quality of life improvements would be created for current and future generations, including a large population of retirees from the baby boom generation.

In the context of business ecology, the sustainability growth cycle described by Parker and Lettes **means creating favorable conditions for planting and growing viable business ecosystems in communities and regions around the world. These business ecosystems would be self-organizing;** that is, they would be guided by shared purposes and values articulated by individual communities, businesses, and regions, and include cyclical life-sustaining flows among participating organizations. This includes creating a demand for sustainable products and services that actually improve the quality of life, while creating optimal value for customers, employees, shareholders, and other stakeholders in the production of sustainable products and services. A win-win-win outcome is created where profits, quality of life, and environmental restoration are self-reinforcing goals of the sustainability growth cycle.

As described in Chapter 1, the U.S. President's Council on Sustainable Development (PCSD), a consortium of business, government, environmental, and community leaders, recommended both subsidy reform and a revenue-neutral tax shift in 1995 to help stimulate sustainable development. More recently, the Industrial Ecology Prosperity Games®, a workshop sponsored by Department of Energy (DOE) National Laboratories in May 1997 in Herndon, Virginia, highlighted several factors that would create a favorable business environment for industrial ecology and sustainable development. Workshop participants included leaders from AT&T, General Motors, Monsanto, Xerox, and other businesses, as well as DOE national laboratories, federal agencies, universities, and nonprofit organizations. In addition to similar tax and subsidy reforms advocated by the PCSD, the workshop attendees highlighted the need for investment support from the financial community, more effective marketing of industrial ecology as a business and development approach, and education of the general public about sustainability. In addition, a high degree of government support (though not defined in detail) and full-cost accounting that establishes a level competitive environment for industry were recommended. Most attendees believed these and other measures would help stimulate sustainable growth cycles in the U.S. and abroad.

THE LIFE CYCLE OF VALUE CREATION

As Chapter 3 describes, value creation is a core process of an organization's metabolism. Value creation is a cyclical, incremental process that begins with an idea or existing products or services, and ends with products or services that have value enhancements that are clearly recognized by customers and other stakeholders. As shown in Figure 2.2, the process involves:

- Creating a formative environment for ideas or innovations
- Planning, research, and development
- Purchasing, acquisition, and outsourcing
- Design and testing
- Production and processing

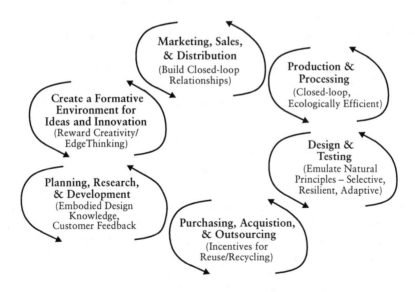

Figure 2.2   The Life Cycle of Value Creation   © 1998 Business Ecology Associates

- Marketing and sales
- Closing the loop with cyclical distribution of products, services, cash, and other life-sustaining flows

Cyclical life-sustaining flows, such as energy, materials, and money, contribute to the value-creation process. The cycle does not stop here, however. Business ecology sees the "leftover" energy and resources as potential new value-creating flows. For instance, residual heated water and steam from a power plant can heat buildings, greenhouses, and aquaculture facilities. In addition, stakeholders, such as customers, employees, shareholders, and suppliers, contribute and/or receive value along the cycle.

Chapter 3 also distinguishes between actual and apparent value creation. Actual value creation improves quality of life for customers and other stakeholders. That is, people experience benefits from products or services that actually meet or exceed their needs, desires, and expectations. In contrast, apparent value creation includes the benefits that we believe we will receive from products and services, which may be greater, lesser than, or equal to actual value creation. Within our industrial, consumption-driven economy, apparent value creation generally exceeds actual value creation. This means customers and other stakeholders do not receive the quality of life improvements that they expect to receive from products and services.

The following discussion focuses on factors that contribute to value creation in a sustainable enterprise and walks through the cycles outlined in Figure 2.2, The Life Cycle of Value Creation. We also highlight:

- Actual value creation, where quality of life benefits are experienced in more enduring, meaningful ways
- Why business ecology stresses marketing and sales that is relationship oriented
- How closed-loop distribution can favorably position your organization in the ecological economy

Completing the cycles of the value-creation process takes time: some estimate that, historically, the average time required to develop an idea into a commercial product can be twenty years. The actual "time to market" varies considerably from technology to technology, and may be as short as several years in fast moving fields like electronics, or even months in the software development field. A general trend is to try to shorten the "time to market," partly because individual inventors or small companies cannot sustain long research and development programs. Even though large companies can better sustain research and development with profits from other product lines, they still face economic pressures that usually work against the innovation and development process. Speed in bringing a product to market usually results in a competitive advantage for the business or organization.

Factors Contributing to Value Creation in a Sustainable Enterprise

In a knowledge-based ecological economy, how is value created? In "Using Knowledge to Create Value" (1996), Tip Parker and Theodore Lettes of Growth Cycle Design, Inc. in Bethesda, Maryland, describe ten interrelated classes of knowledge that create value within organizations and economies:

1. **Market knowledge:** the understanding or vision of products that people will buy at prices significantly higher than the costs of making them. The knowledge may be based on extensive cultural studies, projections of needs far into the future, and learning how to create markets.

2. **Technology:** the scientifically based know-how required to make a product. The Japanese scoured the world in search of technology to use as a cornerstone of their post-war recovery. Much of the technology came from the U.S.

3. **Workforce skills:** the knowledge, dexterity, and abilities of people to make a product. Upgrading worker skills has been a primary goal of the Japanese and their ability to do it well helped them capture much of the U.S. consumer electronics industry.

4. **Manufacturing and distribution knowledge:** the recognition that the process of converting materials into products and getting them to customers around the world is a field of learning that requires continuing study and improvement for each company, plant and product.

5. **Conservation:** knowing how to avoid waste. Much of the Japanese stress on quality is based on their aversion to waste which comes from their dependency on imported materials and fuel.

6. **Embodied knowledge:** information built into a product that adds more to its value than to its cost. Math functions and calendars built into electronic wrist watches are examples.

7. **Knowledge processing:** the ability of a product to increase the value of external information and help create knowledge. Computers, communications equipment, entertainment storage and reproduction devices, measurement instruments, and software are examples.

8. **Design knowledge:** knowing how to use an understanding of aesthetics and human factors in conjunction with market knowledge, technology, manufacturing and distribution capabilities, economics, and other types of knowledge to create successful products.
9. **Relationship knowledge:** the understanding and trust that comes from working with others through networks of enduring relationships instead of just arms-length transactions.
10. **Management knowledge:** the curiosity, reasoning power, conceptual strengths, and leadership abilities that are necessary to develop better ways of combining and integrating the previous nine steps of knowledge to make successful organizations and products.

These insights meld with business ecology's use of natural systems as models for value creation, the innovative design of products and services, and closed-loop product/service cycles. Business ecology can help you capture multiple value-creating cycles by closing the loop on residual flows and by-products. This closed-loop thinking also has significant implications for Parker and Lettes' ten knowledge areas. **In essence, this new ecological economy is being discovered and defined in terms of cycles, efficiency, and reusing resources.**

Business ecology recognizes that information and ideas are a powerful wealth-creating flow within an enterprise. The ultimate goal of sustainable value creation is to use life-sustaining flows, time, and space efficiently to optimize value for stakeholders by creating and delivering high-quality products and services that meet or exceed customer needs and desires while enhancing and sustaining life. What are the specific factors in the value-creation process? These include: creating a formative environment for ideas or innovations; planning, research, and development; purchasing, acquisition, and outsourcing; design and testing; production and processing; marketing and sales; and closed-loop distribution of products, services, cash, and other life-sustaining flows.

GROWING INNOVATION

How does an idea become an invention? How is an invention eventually adopted by people? And how does an innovation penetrate the marketplace, sometimes completely replacing those products or services that came before? Examples of new technologies overwhelming the old include: color televisions replacing black-and-white sets; computers replacing manual typewriters; audio tapes replacing eight-track cassettes; and compact discs replacing phonograph records.

The first phase of the value-creation cycle is the transformation of an idea or innovation into a new product or service. Creating a formative environment for ideas and innovations to emerge in the first place is essential. Such an idea-creating environment is shaped by the core purpose and values, the social DNA, of the inventor or idea champion, the perceived needs and values of potential customers, and the values of the business or organization associated with the inventor or idea champion. Eventually, a concept emerges that gels; the idea or innovation fits in the context of real-world application. Of course, what is considered "real" is affected by social DNA. For example, a society that perceives energy and material resources as limitless, would see little or no value in technologies that encourage energy and material efficiency.

Natural systems provide useful models for understanding how to create a formative environment for an idea or innovation within organizations and their environments. Business ecology applies these models, with emphasis on the following:

- Change happens at the edges
- Being open to innovation
- Why ideas are fragile within organizations
- How ideas survive and develop into commercially viable products and services with market share

Change Happens at the Edges

In nature, change is concentrated along edges. Edges are where life is most pronounced. Growth, for instance, occurs along outer surfaces. Animals congregate along the edges between different habitats, such as forests and open fields. Earthquakes and volcanic activity occur most frequently along the edges of tectonic plates. This moving crust and the biosphere form the outermost "skin" of the Earth. The edges between night and day, seasons, and weather systems also tend to be active zones. Salmon spawn and then die, marking the edge between life cycles, between one generation and the next.

Edge environments, such as the shoreline, can be turbulent, ever-changing, and sometimes violent. Organisms living in such edge zones adapt to their conditions: crashing waves, the regularity of tides, and the intensity of storms. They seek comfort zones, such as pools, and hide beneath rocks and sediment. And so it is with ideas and people. Even when we are close to the edge—close to new ideas and quantum leaps—we seek the haven of comfort zones.

Within organizations, conventional thinking might be thought of as a "center of thinking," a collection of notions that most people regard as standard or status quo—which is far from the edge. In the political world, conventional thinking is mostly moderate and centrist; it is neither left nor right wing, nor is it "fringe." Think of a center of gravity, a balance point where, for example, differently sized partners can balance a playground teeter-totter. A massive partner at a short distance from the balance point may counterbalance a more diminutive partner at a greater distance. So it is with ideas. Many people may share commonly-held beliefs and cluster massively near the center of gravity of conventional thinking, or the comfort zone. "Edge" ideas, held by few, are at a much greater distance from the center of gravity. The entire system of mass and distance, of conventional thinkers and challengers, evolves into a dynamic equilibrium.

Everyone has heard of someone who is regarded as "far out." This expression is often used to describe thinkers and ideas that are far from the norm, or on the edge. Conventional thinking and the "far out" edge represent extreme conditions. One is static, rigid, and resistant to change—yet essential to the creation and the growth of the other, which is dynamic, fluid, and nearly always changing and "improving." Think of an oak tree: the tips of branches change, grow, and flex in the wind, yet could not exist without the support of the trunk. Just as acorns released from these branches yield new trees, new ideas become independent. These illustrations are simplistic, but essentially true. The development of a new religion is another example. Christianity, which has its roots in The Old Testament and Judaism, essentially grew from it to become an independent belief system of its own as proclaimed in the New Testament.

Within businesses and organizations, new ideas become independent, usually through a gradual process that takes them to a zone that lies between the far-out edge ideas and conventional, rooted thinking. This zone might be called the transitional or branching zone. This zone represents ideas that are not too threatening but do, nonetheless, represent a new way of seeing. Many successful inventions are found in this zone because market acceptance of a potential product does not require too far of a stretch. In the oak tree analogy, these inventions are branches growing close to the tree's trunk, not the outermost edges where the twigs, leaves, and acorns grow. These are the ideas and innovations that often make it through the value-creating phases of research and development and design and testing.

In the next value-creation phase, planning, research, and development, the idea undergoes a series of feasibility checks that are made to ensure that continued development is wise. Market feasibility ideally is defined by direct contact with customers and builds on this mutually beneficial relationship. Focus groups and product samples are well-known examples of ways to get customer feedback. Technical feasibility is often determined by an engineer or a scientist to ensure that the invention can actually work, and that it can be built with known materials. Economic feasibility involves estimating costs and benefits, with an adequate margin for profitability. These and other feasibility checks are repeated often during the process of developing a product or service to determine whether development should be continued or abandoned. Again, values or social DNA may sway the result, especially with respect to economic and market feasibility. The value of an energy efficiency device, such as a compact fluorescent bulb, is not fully recognized in an economy that does not appreciate the full cost of producing and distributing energy.

If an idea is found to be viable, various options are considered to move a project or enterprise forward. Clearly, purchase, acquisition, and outsourcing decisions are values-driven and can have far-reaching impacts on an organization's stakeholders and its value-creation process. Values such as trust, quality, reliability, and life-cycle thinking can profoundly shape how a business or industry develops and acquires resources, processes, and capabilities.

Carl Henn, a leader renowned for bringing ecological principles to accounting, investment, and other business professions, points to the prominent role of the U.S. military in stimulating change with life-cycle thinking, systems and planning tools, and the power of purchasing. According to Henn, modern business practices and tools such as life-cycle assessment, full-cost accounting, and scenario planning evolved from military applications (e.g., logistics research for aircraft). Henn suggests that the military through its purchasing, acquisition, and outsourcing policies brought about significant changes within its stakeholder community (e.g., quality specifications for weapon systems components). Similarly, he believes that industries can leverage change by exerting their purchasing power with their stakeholder communities (e.g., suppliers are required to use recycled materials in their products). Henn believes that such **purchasing, acquisition, and outsourcing decisions are strategic opportunities for stimulating sustainable enterprise and closed-loop business practices.** Here's an example from Robert Steutiville's *In Business* interview with Michael Smith, an AT&T purchasing manager:

AT&T, which purchases more than 100 million pounds of paper annually, has made significant strides in using recycled content. [In 1993],
the company used recycled content in 100 percent of its direct mail.
"The important point was that it didn't raise our costs," Smith said.
"We were able to find a good source of paper at a reasonable cost."
    Suppliers need to be made accountable to buy recycled programs,
Smith stressed. "Let suppliers know you are looking for recycled products—these people will find them."

As this example illustrates, the point where purchasing, acquisition,
and outsourcing decisions are made is a strategic opportunity for encouraging ecological efficiency. The next step is extensive design and testing to
improve or develop components in a laboratory. The laboratory effort
eventually produces either failure or "proof of concept" at a bench- or
laboratory-scale. If successful, the "proof of concept" tests may produce
either marginal or convincing results. If convincing, a next step is to develop an actual working prototype. Lessons learned throughout the development process usually are incorporated in successive prototypes before a
pilot-scale version is developed. With success at pilot scale, a full-scale demonstration may be considered. Full-scale application not only demonstrates
that a product or service works at commercial scale, but also builds con-
fidence in those who are likely to first adopt the product or service, the
"early adopters."
    During these demonstration phases, much attention may be given to
the details of production and processing, with a goal of reducing production costs when the innovation is produced in large quantity. Preproduction
engineering often results in subtle adjustments to the design and possible
substitution of construction materials. A preproduction version may be
subject to limited production runs to test the viability of production techniques and the market before funds are committed to mass production. **In
the new ecological economy, an important aspect of preproduction adjustments includes planning for the future in terms of resources, by-products,
and demand.**
    For example, imagine that you are a manufacturer of standard grey
office chairs, Model 1945-D2. What does a manufacturer do about predic-
tions that the market will dry up? Produce fewer chairs? Find lucrative
ways to recycle old chairs? Downsize? Test the market with different products?
    Suppose there are well-supported proposals for a line of civilian chairs
suitable for many uses and styled with designer colors, recycled woods, and

fabrics. These proposals are still in the zone of conventional thinking because the basic assumptions and solutions have not changed.

On the edge of this zone is the transitional or branching zone. In the example of the chair, ideas in this zone call into question the specialized nature of "chairness" and include new concepts of chairs, such as "beanbag" and "ergonomic" chairs. If change is imminent, why not rethink all the options, including whether or not producing chairs will be a wise move in the long run?

In the emerging ecological economy, **this type of thinking that branches away from convention, and even far-out edge thinking, is important because it widens the window for innovation.** The problems and limitations businesses and organizations are facing today demand innovations—ideas and solutions—that meet customers' needs and take into account future resources and by-products. Monsanto's NewLeaf™ potato, for instance, is a match of innovative business-ecology thinking and resource conservation. This potato, which has genetically encoded information to protect it from viruses and pests, does not need to be sprayed with financially and environmentally expensive toxic pesticides. Another example is from Malden Mills, where researchers have redefined "fabric" and introduced in 1993 the Polartec® Recycled Series of fabrics. Instead of using virgin polyester fibers, these Polartec® fabrics are made from recycled PET (polyethylene terephthalate) plastic containers, acquired from Wellman, Inc., one of the world's largest plastic recyclers.

In these examples, both Monsanto and Malden Mills have kept a keen eye on their customers, their other stakeholders, and the environment. Their innovations, which sprang out of a values-based commitment to reduce the environmental impact of their products, are examples of thinking in the transitional or branching zone. So is Monsanto's and Malden Mill's resourceful, creative questioning of traditional solutions. The result? Sustainable ideas become profitable products that are readily accepted in the marketplace.

Figure 2.3 shows the many perilous pathways between idea and market acceptance. The shaded figure on the left represents different zones of thinking within a business or organization. At the core is conventional thinking, the status quo. The next layer away from the core represents the transitional or branching zone. This zone represents ideas that bridge conventional thinking and assumptions with innovation. The outer zone represents the domain of the most radical ideas, those edge ideas furthest from conventional thinking.

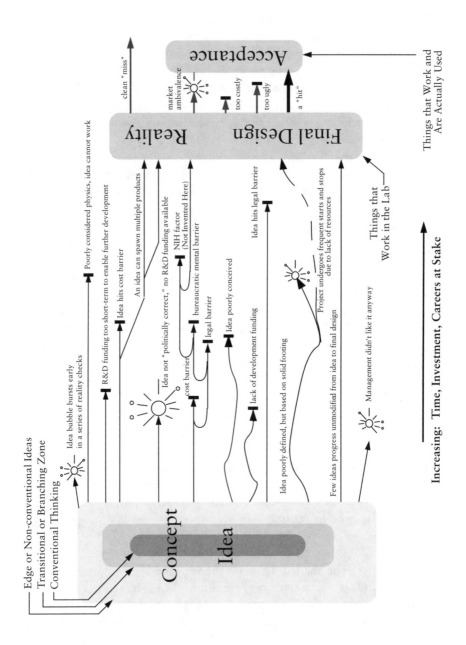

Figure 2.3   The Perilous Path from Idea to Market Acceptance

The Fragility of Ideas

Ideas face many challenges and tests on their fragile voyage from concept to market acceptance and eventual market share. Most do not make it. An analogy to reproduction is striking. Sea turtles release thousands of eggs into the environment. Many are not fertilized. Of the ones that are fertilized, most are eaten within a short time. As the remaining eggs eventually grow into small turtles, the hostilities continue with devastating attacks from predators in the air. The odds for survival increase only when the turtles reach a critical size. Surely, **Darwin's observations regarding the survival over time of the "fittest" applies as well to human ideas.**

An example of the survival odds for new ideas may be found in the Energy-Related Inventions Program operated by the U.S. Departments of Energy and Commerce. This small grant program, established in 1974, is designed to help private inventors evaluate and commercialize their inventions. Of every 100 proposals that are received by the government, about 50 percent are rejected at first-level screening. At a second-level screening, about 90 percent of those remaining are rejected, leaving only 5 percent of the original pool of proposals. Of the 5 percent of the proposals recommended for funding, only about 2-3 percent actually receive funding. Of the inventions receiving funding, about half are successful in the marketplace.

Figure 2.3 illustrates some of the hazards facing an idea or a concept. There are the harsh breezes of physics that provide reality checks that cause many idea bubbles to quickly burst. Sometimes the concept requires too much energy to drive toward a successful conclusion. Sometimes a seemingly great idea will die due to neglect or lack of adequate funding. Some ideas can work in theory, but cannot make the jump to reality. Some ideas can be made to work in the lab, and yet fall short of the durability and robustness required by the market. In addition to the problems inherent in developing any new technology, ideas often face the overriding management desire for the mean time-to-market to be measured in days, preferably minutes. Many organizations do not have the stomach, the will, or the patience to develop a technology to a state-of-market worthiness. Those that do often find their beloved offspring stolen by pirates. Figure 2.3 also shows that apparent dead ends can actually spawn new ideas and new products. When final designs are finally functional in a laboratory setting, there are still further hurdles that separate those things that work in the lab from those things that work in the lab and are actually used in the marketplace.

Following are the key factors affecting product and service design and development, as presented in Figure 2.3:

1. Any design, including the design of organizations and technologies, is a visualized concept. Social DNA (i.e., core purpose and values), because it shapes perspective and behavior, plays a critical role in the conceptualization and innovation process.

2. Design and development options are always available and each may have value. Again, the design and development environment is shaped by social values. Thus, final outcomes are influenced by the prevailing values of an organization and its environment.

3. Multiple design options are valid and are often preferred to ensure overall success of a market "hit." Nature often experiments, through trial and error, with many different designs until a successful one emerges. For every successful organism, many more have perished.

4. Each choice taken along each design and development pathway increasingly excludes other possible approaches. In human inventions, as in natural systems, overspecialization and poor adaptability can lead to extinction, unless there is a clear and enduring need for the product and service.

5. Some efforts will not bear fruit. Recognize that few ideas go from concept to reality along the pathway originally conceived. Evolution and the reproduction of organisms, for instance, often include more dead ends than survivors. Nature often stacks the numbers in its favor or develops a particularly robust, resilient, and adaptive design.

6. Nature provides for the full spectrum of organisms that are necessary—from the highly specialized to the "generalist"—to create systemic balance. Similarly, an organization needs to have an understanding of the entire market in which the product or service being designed will fit.

7. Some approaches will appear successful at first, but may fade from preference as new ideas or knowledge is gained, and particularly if another design is more effective in meeting market needs. This is OK. The environment is always changing and uncertain; timing can be critical and surprises will occur.

8. Like survival of the fittest in nature, the best ideas and designs often prevail. Designers and managers should welcome further investigation and competition among ideas and designs.

9. Adherence to dogma and old product lines deprive the market of fresh new designs and products, and simultaneously deprive an organiza-

tion of viable, relevant opportunities. Knowing when to adapt is vital in a changing environment.

10. Relationships with stakeholders, like relationships among organisms in nature, build robust, resilient, and adaptive environments for innovation and development. Direct contact with customers, suppliers, and other stakeholders along the value-creation process helps ensure the viability of new products and services.

Being Open to Innovation

Being open to innovation strengthens all phases of value creation. There are a number of ways of capturing opportunities presented by edge-related innovation. But first, *will you know the innovation when you see it?* Are there people within your organization that are marginalized because they have a different perspective? Does your organization appreciate and nurture a diversity of perspectives and approaches? Swiss watch manufacturers, for instance, developed the first electronic watches, but were unable to embrace the paradigm shift fast enough. Research and development personnel were unable to convince management to shift from their traditional spring-wound, geared time pieces. The result: Japanese manufacturers, who saw the opportunity of this innovation, developed and dominated the electronic watch market.

Second, *look to the edges of disciplines* to see fundamental change and innovation, especially where disciplines overlap. In the 1930s, for example, hydrologist Charles Theis observed approaches used by engineers to understand heat and electrical flow, and deduced that a similar set of forces and relationships drove the subsurface movement of ground water within water-bearing geologic formations called aquifers. At first his colleagues thought his idea was preposterous. Today, Theis's work is considered one of the major breakthroughs leading to the development of modern ground-water hydrology. His methods are used by water professionals around the world to characterize and manage ground-water resources.

Third, *does your organization allow the diffusion of innovation from the outside?* Are you open to ideas from other types of organizations and cultures? Or does the social DNA of your organization build an impervious membrane to this diffusion? For instance, modern Western medicine, which tends to favor technological solutions to sickness, is being challenged fundamentally by holistic approaches that focus on wellness, diet, and the interconnectedness of mind, body, and spirit. Much of this thinking is coming from the Eastern world and indigenous peoples, where technological solutions are less available. In the past, these approaches have been marginalized within the Western medical community. Today, a better informed public and portions of the medical community are demanding that alternatives, such as acupuncture and herbal and dietary remedies, be adopted when they are proven to work.

Fourth, *nature provides an infinite pool of inspiration and innovation* if one is able to adopt an ecological lens rather than only a mechanistic point of view. Nature certainly has features that appear to be mechanistic, such as the reaction steps in cellular metabolism. But using a mechanistic lens exclusively shuts out other dimensions of how nature works. It is like seeing the world in black and white when it is in full color. The fundamental aeronautical innovations of the Wright brothers, for example, were inspired by watching birds in flight, especially how their wings were shaped and used. But as noted in Chapter 3, the elegance and wonder of a bird go well beyond its aeronautical design.

SURVIVING IN THE MARKET ENVIRONMENT

We know that the marketplace, like the natural environment, is both uncertain and predictable, full of opportunities and dangers. The diffusion of new products and services, or new enterprise, is analogous to new plants and animals attempting to carve out a niche in a resource-limited ecosystem—they compete for time, space, and other life-sustaining flows with already established organisms as well as recent pioneering species. However, by closely observing repeating patterns, such as seasons, one can predict which species—like ideas—will make it. Successfully transforming an idea into a tangible product in the marketplace is an inexact science at best. However, certain observations from nature do apply.

For example, S-curves, which are used to describe market acceptance of a new technology or product, are also used in biology to describe population growth in a resource-limited environment. Figure 2.4 shows the diffusion of a product, service, or enterprise into a competitive market. Much like organisms competing for limited resources within an ecosystem, new products, services, and enterprises strive to carve out niches that build on their competitive strengths. Those unable to adapt to market changes often perish.

Nature experiments all the time with new designs and prototypes, but not all species succeed or endure. Some, such as the horseshoe crab, have endured for millions of years. Other, less fortunate species end up in fossil

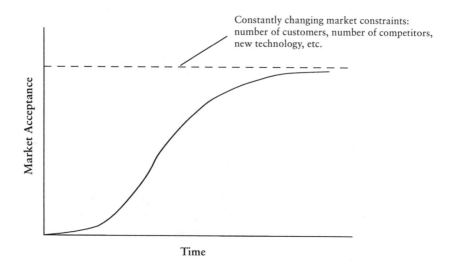

Figure 2.4   Diffusion of Products, Services, or Enterprises into the Marketplace
© 1998 Business Ecology Associates

records, or more recently our museums, zoos, and protected habitats. Why? Successful organisms—and organizations—reflect robust ecological design and adaptive instincts. Nature equips its organisms with *strong relationships with other organisms,* such as what exists between woodpeckers and trees. Woodpeckers eat insects from the bark of trees and use tree hollows for nests. Trees, in turn, benefit from woodpeckers by being protected from harmful insects and disease. Nature also equips its organisms with numerous *feedback loops to sense and interact* with the environment and provide the means for adapting to changing conditions. The homing instincts of migratory birds is a good example. Year after year, they find their way between their cold- and warm-weather habitats. In addition to this *embodied design knowledge expressed in DNA,* organisms also develop social organizations within and among species. These natural systems have coevolved dynamically with their surrounding environment.

Natural systems are *resilient and adaptive* because they pass success secrets and lessons learned from generation to generation. They draw from a diverse pool of knowledge, including genes and instincts and recent adaptations, to current conditions, such as learned behavior. Especially adaptive species, such as humans, cockroaches, and coyotes, have broad-ranging habitats and diverse sources of food and sustenance. This is similar, for instance, to how a larger company may survive by marketing a broad variety of related products.

It is no surprise, then, that emulating natural systems can give your business or organization a natural edge. In fact, organically designed organizations are strategically positioned for adaptation and innovation. As pointed out by Sue Hall and Bryan Thomlison in Chapter 6, companies, such as Arm & Hammer and Shaman, for instance, have experienced significant bottom-line benefits by creating healthy and open relationships with their stakeholder community. This diffusion of ideas and market knowledge from outside can keep an organization in sync with changes in its environment, helping it adapt to new opportunities and challenges.

The fact that **business ecology is relationship-oriented is another factor that gives organically designed organizations a marketing edge.** Traditional, consumer-based marketing, distribution, and sales is object-oriented. It makes people "vessels of consumption" to be filled with "things" that the economic machine produces. With business ecology, the customer is recognized as an essential, "sustaining" stakeholder within an organization's environment. Thriving businesses view their "customers" as vital to their success, and cultivate lasting, respectful relationships with customers and other stakeholders.

Business ecology recognizes that a satisfied, even delighted, customer represents a sustained cyclical earnings stream and can be a highly effective "marketing agent." Marketing establishes who those customers are, what and how their needs are changing, and develops strategies for keeping those customers satisfied with products and services. A product or service may not sell or be accepted for a number of reasons. **As common sense would dictate, most of these variables point to the relationship with the customer.** The product or service will sell steadily if customers recognize it as meeting their needs or desires, and the market is not already saturated by tough competitors, and the customer population is not too limited.

Being relationship-oriented, business ecology not only recognizes that customers and other stakeholders are intrinsically valuable—even vital—to your company, but also recognizes that your business or organization is part of a larger web of relationships. This web is developed through employees, advertising, public relations, professional partnerships, and community relations. Developing healthy stakeholder relations, as the examples in Chapter 6 illustrate, is essential to your company's bottom line and long-term viability. Total quality management and other management approaches have rejuvenated this simple common sense that has been obscured in a climate of mass-production, consumption, and planned obsolescence.

A critical part of your organization's relationship web are distribution loops. These loops convey products and services among an organization and its customers, including, for example, the return, exchange, or repair of products. Compensation for the provision of products and services can be the transfer of money (selling), services (barter), or approval through continued funding support, in the case of nonprofit and government organizations.

Sales ensure that the desired transactions regarding products and services take place, with a focus on increasing profits and profit margin. The value-added products and services that are created are exchanged for money and/or other life-sustaining flows. Bartering services and resources is an example of compensation that involves little or no cash flow. Nonprofit and government organizations, for example, must "sell" their programs to funding entities, with a focus on increasing funds and increasing fund-raising efficiency.

Business ecology radically redefines marketing, distribution, and sales in the context of sustainable development, sustainable enterprise, the ecological economy, and emerging business trends, such as eco-accounting, building partnerships, and strong community relations. Marketing, sales, and distribution systems in a closed-loop economy link supply and distribution, and interact dynamically with a web of relationships with customers and other stakeholders. Like the circulatory and respiratory systems in our bodies, marketing, distribution, and sales loops carry life-sustaining flows. These include metabolic inputs (such as food, cash, energy, and materials), metabolic outputs (such as products, services, and by-products), and residual flows (such as energy, water, and materials). These closed loops are also like the feedback loops that animals use to sense and interact with their environment. The closed loops used by one organism (or organization) are metabolic products, services, and by-products from other organisms (or organizations). Chapters 4 and 5 provide numerous examples of how such "closing-the-loop" behavior is good for the bottom line.

For instance, Mitsubishi Electric America plans to develop several "take-back centers" to close the loop on their electronic products. This will strengthen their relationship with environmentally conscious stakeholders, including customers, suppliers, and shareholders. Building a complete business ecosystem by closing the loop will help Mitsubishi save money through ecological efficiency and create new areas for sustainable enterprise, such as demanufacturing and remanufacturing of electronics products. It will also position them favorably in the closed-loop ecological economy.

Marketing, distribution, and sales loops carry life-sustaining flows back and forth from customers to value-creating systems where the resources are transformed into marketable, value-added products and services. In this sense, value-creating systems are the suppliers. The customers include all the stakeholders who receive value from suppliers. Once created, products and services are either used in the internal economy by internal customers, or are carried to the external environment where they are exchanged for money and other life-sustaining flows from external customers. This is not unlike an estuary in nature, where tides and currents cause a productive mingling of fresh water, nutrients, detritus, oxygen, and salt water to yield an abundance of animals, aquatic plant life, and wetlands, which in turn filter, decompose, and recycle residuals from other organisms and parts of the estuary. Similarly, "used" products and services that no longer meet the customers' needs and by-products become residual flows. These flows from the external environment are filtered in and metabolized by the business organism, creating a new value-creation cycle. "Take-back legislation" in Europe and the development of ISO 14000 standards, which were discussed earlier in this chapter, are examples of trends that are stimulating these continuing cycles of value creation and other closed-loop behavior within and among companies and their environment.

Here is a summary of salient points for each phase of the value-creation process:

Create a formative environment for ideas and innovation
—Make time for and reward creative "edge" thinking and innovation
—Change happens at the edges

Planning, research, and development
—Market feasibility defined by direct contact with customer
—Embodied design knowledge/life-cycle thinking
—Values-based planning directed towards achieving a shared vision

Purchasing, acquisition, and outsourcing
—Create incentives for reuse and recycling

Design and testing
—Emulate natural principles—selective, resilient, and adaptive—to refine the best ideas

Production and processing
—Speed to the marketplace and efficient use of resources
(just-in-time, just-in-place)

Marketing, sales, and distribution of products, services, cash,
and other life-sustaining flows
—Good relationships build short- and long-term profits
—Build closed-loop relationships, cyclical thinking, feedback loops
—New value-creation cycles started from by-products and residual
flows

The life cycle of value creation is a synthesis of ideas, relationships, and inspired management based on the core purpose and values, the social DNA, of a business or organization. How efficiently does your organization use life-sustaining flows, time, and space to optimize value creation? In this sense, your business or organization's value creation, and ultimately its success, equals the extent to which it creates an improved quality of life for all its stakeholders.

The life cycle of value creation is an inner circle, a part of the larger design of your business or organization, as discussed in the first part of this chapter. At the root of successful value creation is a design environment and organizational model that clearly defines its core, underlying values. These values (i.e., social DNA) have been deeply woven into our worldview and reflect the passing down, through the centuries, of beliefs and priorities borrowed from traditional and ancient cultures. Some of the wisdom from these societies is particularly relevant today, including a respect for natural rhythms; collective cooperation; a holistic view of mind, body, and spirit; and the strong organizing forces of belief systems, shared rituals, and traditions. In the new ecological economy, viable, adaptive organizations will adopt such traditional wisdom along with organizational models that emulate natural systems.

The model of business ecology, its mindware, synthesizes centuries of cultural wisdom with a close observation of natural systems. In the context of value creation, one of the key principles of natural systems discussed in this chapter is interdependence—thriving organisms and organizations are not isolated. In fact, successful businesses and organizations selectively filter in and out a rich variety of information, individuals, and ideas. Growing, communicating, and being a part of a larger web of relationships and responsibilities are all essential to your business or organization's resiliency and adaptability. For your business or organization to branch into sustainable, innovative patterns and relationships, a clear definition—and in some cases, a new definition—of your values, perspective, and behavior is needed. Business ecology shows you how your organization really works by allowing you to see these and other essential factors that shape your organization's success.

3

The Business Organism

We are all controlled by the world in which we live, and part
of that world has been and will be constructed by men. The
question is this: are we to be controlled by accidents, by
tyrants, or by ourselves in effective cultural design?

B. F. *Skinner*

Organizations exist throughout nature. Wolf packs, flocks of birds, bee-
hives, and the cells in our own bodies are elegant, awe-inspiring expressions
of "community." These natural systems have coevolved dynamically with
their surrounding environment. They are a product of ancient, evolutionary
history, determined in part by genes and instincts, as well as recent adaptations to current conditions, such as learned behavior. Such natural systems,
because they pass success secrets and lessons learned from generation to
generation, are resilient and adaptive. They can even anticipate change.
Squirrels and bears, for example, sense and respond instinctively to the coming of winter and the return of spring. We have much to learn from this
exquisite organizational wisdom that has evolved over 3.5 billion years.

In contrast to these life-sustaining organizations, many human organizations suppress our deeper connection to ourselves and our environment. They may also stifle creativity, innovation, imagination, intuition,
and personal growth. They are rigidly hierarchical, reflecting the features
of the mechanistic-industrial paradigm. The design and operation of such
organizations reflects a profound separation from nature and, thus, ourselves. Such organizations often cannot flex and adapt quickly to change;
in fact, these institutions fear change, even though it is one of the most
basic characteristics of life.

Developing dynamic, life-sustaining organizations that are resilient and adaptive requires, as Peter Senge suggests, that we overcome our learning disabilities. Central to this learning process is rediscovering our relationship with nature. The notion that we are separate from and can control nature without affecting our own destiny is an illusion. Once we recognize this, we can shed our "mechanistic blinders" and perhaps have the humility to see ourselves as part of the web of life, rather than its "masters." When we do this and use natural systems as our models for organizations, profound, enduring change is possible.

SEEING BUSINESS WITH NEW EYES

We use mental models all the time, often without realizing it. Peter Senge defines mental models in *The Fifth Discipline* as: "deeply held internal images of how the world works, images that limit us to familiar ways of thinking and acting." Models or paradigms help us communicate, interpret, and associate information, facts, and ideas. They shape our perception of reality. The incident described below illustrates how powerful mental models can be. Author Dave Bassett recalls:

> In my youth, I spent a great deal of time in the forest, camping in many favorite wooded haunts. Without realizing it, I learned the language of the forest and some of its creatures. I understood their behavior and could anticipate their next moves. In this context, I understood the essence of "birdness" as a naturalist might. I could distinguish among them as a biologist might. I took winged flight somewhat for granted as a "bird thing."
>
> In college, physics and engineering demanded a different view: not one of taxonomy and distinction, but of mathematical analysis, of gravity, of vectors and force distribution. Time went by, and soon the idea of looking at objects as representations of force fields became comfortable. Physics and engineering provided a new lens through which to look. Things were "things" after all, and represented distributions of energy and mass—all subject to the laws of physics.
>
> One day during spring break, I took a walk in the woods to visit some of my favorite haunts, to commune with the trees and to watch the woods come alive with the first signs of spring. Deep in the woods, I came upon a special place where three properties joined. The place was special partly because it represented a transition between open fields and deep woods. It also represented the juxtaposition of two systems, one natural and one man-made. The natural world saw only

habitat and ecosystem, the blending of open field into scrub brush, and then into old growth forest. The man-made legal world saw only three invisible lines intersecting and conveying property rights to individual owners. The creatures of the forest simply did not exist in the legal world, only man.

Snow was melting, and a warm breeze from the southwest was bringing smells and hopes of spring to the forest. A sparrow flew into this peaceful scene and sat upon a branch, studying the intruder. And the intruder studied back. What distribution of forces was required for steady flight? For intermittent, twisting, and sporadic flight? Where was the center of mass? How did the center of mass change with each flap of the wings? What were the lifting surfaces? How much of the wing actually provided lift? How much provided control? What was the control logic to determine whether energy flowed into lift or into control? What did the flowchart look like? Where in the logic flowchart would you place, should you place, "system overrides" that would take priority control in the event of danger or if evasive action were required? How was wing shape modified during a flap cycle? How did energy flow between food input and flapping output? What percentage of input energy was required for motion? How much for heat? How much to run the control system? Should an analytic three-dimensional coordinate system be fixed with respect to the environment, or fixed with respect to the moving bird? If with the bird, should the X-axis be aligned with the bird's longitudinal axis every instant, or should it be time-averaged? If so, over what time period? And only during flight, or also while landing and sitting? And this landing business looked complicated. What was the interplay of diminishing lift force and inertial force during a landing? What if . . .

Suddenly, without warning, the lens shattered. What on earth was happening? The "engineering lens" provided many interesting questions, yet it failed utterly to capture the seamless elegance of life and motion. During the brief moments when a thousand questions flashed, the bird became a "thing." Not one of nature's special creatures resplendent in its own magnificence, but a "thing" to be studied, analyzed, understood. What was missing was a sense of awe and dumb wonder; a sense of connectedness between viewer and bird, and between bird and all other living things. What was missing was respect for the bird's space, its own special places in the air and on the branches. Reverence and respect were missing totally from the analytic framework.

At this instant, I found the "engineering lens" repulsive. What it took away was far more than what it provided. This realization was very troubling. It implied that a career path, while possibly economically lucrative, was spiritually bankrupt. Later in life, I would understand that these seemingly contradictory views were not mutually exclusive. That is, both views were valid; each view did not invalidate the other. One can learn to turn lenses "on" and "off" at will. A professor of meteorology would later advise that seeing through an analytic lens could only enhance the beauty of the natural world and heighten appreciation of it. While probably true, the evidence suggests that for most people, once a new lens is fashioned, it is used exclusively. But much can be gained by taking back control of our own minds. By exercising our "active choice," we can choose to turn them "on" and "off." We can also recognize and use each lens to see the world "as it is" from these different perspectives, to gain insights about the elegant design, "wholeness" and beauty of creation.

What is the valid model for organizations? The lens of the industrial-age worldview is fast becoming obsolete. In their book, *A Simpler Way*, Margaret Wheatley and Myron Kellnor-Rogers describe this obsolescence and their search for new organizational models:

> The mechanistic image of the world is a very deep image, planted at subterranean depths in most of us. But it doesn't help us any longer. Our own search for new ways of understanding has led us to philosophers, scientists, poets, novelists, spiritual teachers, colleagues, audiences, and each other. We keep exploring what we can see when we look at life and organizations using different images.

How do we shed our "mechanistic blinders" to see business with new eyes? Dee Hock suggests creativity happens when we remove the clutter and noise from our minds. **Creativity fills the void.** Thinking about organizations and ourselves in new ways requires that we do this type of mental house-cleaning. We must look closely at the accepted ways of doing things and ask: Do they work? If not, why? Are they sustainable? What is the worldview, the set of underlying assumptions, upon which they are based? Let's look briefly at the underlying values and worldview that shaped Western, industrialized culture, and some common metaphors and models used to describe business and the economy.

In the 1930s, the advertising industry in America created the "consumer." Since that time, it has not been enough to meet existing customers' needs. "Wants," disguised as "needs and personal fulfillment," have been actively cultivated by the commercial media, which reach even the very young. Somewhere along the way people stopped being customers and citizens, and became addicted consumers. In this context, the industrial production system views people, communities, and the environment as expendable, replaceable, and inexhaustible resources to exploit. Consumption, after all, fuels expansion. The Home Box Office (HBO) film, *Earth and the American Dream*, directed by Bill Couturié (1992), describes the price the Earth and its inhabitants have paid in fueling the industrial, consumption-driven economic machine. Here are two provocative quotes from the film that expose the underlying values of the industrial age:

> The final victory of man's machinery over nature's materials is the next logical process in evolution. Machinery, science, and intelligence moving on the face of the Earth may well affect it as the elements do—abuilding, obliterating, and creating. And they are man's forces . . . and they will be used to hasten his dominion over nature. And each gain upon nature, adds to the quantity of goods to be consumed by society.
>
> *Simon Patten, political economist*

> Our enormously productive economy demands that we make consumption our way of life, that we convert the buying and use of goods into rituals, that we seek our spiritual satisfaction—our ego satisfaction—in consumption.
>
> *Victor Lebow, marketing consultant*

Today, we continue to be shaped in subtle and not so subtle ways by the industrial-age worldview. The economy is often called the "engine of growth." Terms such as efficiency, performance, production, productivity, control, and management—which represent valuable and important functions—are often used for companies and other organizations. They are also qualities associated with machines. Advertising media frequently depict businesses as lock-step, synchronized, omnipotent, and lightning-fast. In our Western worldview, financial markets, corporations, technologies, CEOs, and even celebrities often exhibit god-like qualities, with powers that seem to control or defy the laws of nature. These and other images of eternal wealth, power, and youth form the mythology of what is largely an industrial, commercial, and entertainment culture. It is the desire to "touch these gods" that keeps many consumers buying goods and services that promise such nature-defying fantasies as eternal success, youth, beauty, and power.

Today, the lines blur between what are consumption fantasies and those purchases that are realistic, obtainable, useful, and truly enjoyable.

As Theodore Roszak, Director of the Ecopsychology Institute at California State University at Hayward, California, and editor of *Ecopsychology: Restoring the Earth, Healing the Mind* has written:

> Many people go to shopping malls because they find some diversion from their own inner problems or they seek to prove themselves through acquisition. Yet, they're aware of the fact, I've discovered, that these are feeble, ill-considered ways of trying to meet needs in their lives. You often find that the bad environmental habits of people are connected to thoroughly legitimate human aspirations, but they've been diverted into the marketplace, into getting and spending, and into production and consumption.

In Roszak's 1995 interview with *Common Boundary* magazine, he also noted,

> There are plenty of human beings in the indigenous cultures who don't share our bad habits, so you might insist that, psychologically speaking, the human psycho-gene pool still includes all these other nonindustrial people who have not gone the way we've gone. Can we learn from people who have never taken much stock in industrial progress?

It is difficult, however, to break free of our consumptive lifestyles. We are bombarded daily by the message that "getting rich" and having one's every material need met will bring personal and spiritual fulfillment and make us more important and attractive to others. Paying cash or credit for a staggering array of goods and services is glorified as "living the good life." In fact, consumption *has* become a way of life, a national hobby, which supports the growth of malls and huge chains that are eclipsing vibrant, diverse communities and neighborhoods where people once worked, worshipped, socialized, and shopped together.

In our industrialized society, commercialism is not the only thing which shapes us. Technologies also mold our lives, including how we relate to each other, how we perceive ourselves and nature, and even the pace at which we live. Think of how the automobile, television, telephone, video games, and the Internet have influenced how we perceive and experience the world. While these technologies have brought positive things into our lives, such as conveniences, fun, speed, mobility, access, and novelty, these same technologies can shape how we think, perceive reality, and even interact with each other. When they mold our values in a negative way, or detract from our time to enjoy those simple, meaningful life pleasures that connect us to each other and the natural world, they can erode our quality of life, our "humanness." Many groups, such as Computer Professionals for Social Responsibility, the Society for Research and Initiatives for Sustainable Technologies and Institutions, and the Lion and the Lamb Project in Bethesda, Maryland (which offers alternatives to the violent media to which children are exposed), are seeking a more balanced use of technologies to ensure that our lives, families, communities, and environment are enhanced—not degraded. Such grassroots groups often use technologies, such as the Internet, to reach and educate their constituencies, and develop strategies and projects to promote more appropriate uses of technologies.

The terms that describe the qualities of living systems can also apply to business, the economy, and technologies. These include: "robust," "sensitive," "growing, "changing," "cyclical," and "healthy." It is not unusual for these organic qualifiers to be mixed with mechanistic terms. For example, how can the economy be an "engine," and at the same time be "healthy" and "growing"? Could we ever see the "economy" as a garden? Our employees as "seeds"? Companies as living? Perhaps this mixing of metaphors marks the transition in thinking between the old and new organizational paradigms.

Recognizing the power of these deeply ingrained images, how can you transform your organization? Chris Turner is the chief "learning person" within Xerox Business Services, a fast-growing, highly communicative offspring of Xerox, the copying and document management company. Here is what Turner has to say about organizational models and cultural transformation during an interview with *Fast Company* magazine in 1996:

> You start by changing how you think about the organization. We've grown up in an old-fashioned corporation that's organized functionally. People think of that kind of a company as a machine. That's how they talk about it: "We're a well-oiled machine." It's mechanistic, reductionistic language. It tells people they're just cogs in the machine.
>
> I think of our organization as a dynamic, living system, like an environment. You can't treat a natural system like a machine. All you can do is create experiences that disturb a natural system, and then it decides how to respond.
>
> My job is to disturb the system. I give people new ways to think. It's more a matter of offering people different perspectives and influencing their thinking than trying to drive them.

Certainly Turner embodies the term organizational "change agent," those individuals within an organization who cultivate innovation and life.

AN ORGANIC MODEL FOR ORGANIZATIONS

Figure 3.1 is an organic model for organizations. Life-sustaining flows are dynamically exchanged both within the business organism and between the organism and its environment. Chapter 4 looks more closely at each of these close-looped flow systems and shows how this comprehensive, systemic perspective improves your organization's viability. These life-sustaining flows, together with the stakeholder community, determine your company's "environment," including its metabolism, niche, and habitat.

Business ecology provides a powerful lens for seeing your organization's potential and the vital relationships that sustain it. It blends proven strategies with the organizing elegance of natural systems. Several examples in this chapter demonstrate why the business ecology paradigm works. These examples also demonstrate how some leading companies are already successfully applying, to varying degrees, the principles of business ecology.

We will first look at metabolism, your company's internal economy, and the three vital processes that comprise it: values-based organization and development; value creation; and cyclical flows. This simple, elegant organizational model, which includes more specific areas such as research, product development, purchasing, marketing, and human resources, can be applied to organizations of every scale and type. Next, we will examine niche and habitat, with a few examples. Finally, we will apply business ecology to a specific industry—aquaculture—to illustrate the power and

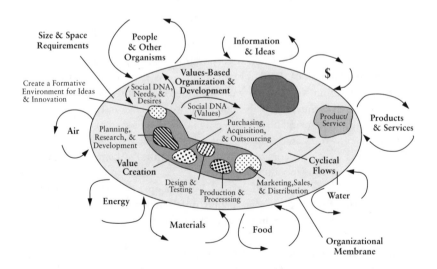

Figure 3.1   The Business Organism © 1998 Business Ecology Associates.

usefulness of the business ecology model. The insights introduced in this chapter and developed more fully in subsequent chapters, will help you strategically position your company in today's rapidly changing business environment.

METABOLISM: YOUR COMPANY'S INTERNAL ECONOMY

Metabolism is the sum of the core processes that support a company's internal economy. As suggested by Figure 3.1, the cell provides a wonderful model for seeing your company as a living system. Here's how Joseph Levine and Kenneth Miller describe cellular metabolism in their textbook, *Biology: Discovering Life*:

> We refer to the sum of all chemical and physical reactions associated with life as *metabolism*. In a way, the metabolism of a cell is nothing more than its internal biochemical economy. The economy of a cell is every bit as varied and complex as the economy of a nation, and studying the metabolism of a cell is just as challenging.

According to Levine and Miller, there are two types of metabolic pathways within cells: (1) processes that combine smaller things into larger things, and (2) processes that break up larger things into smaller things. Within living systems, processes that create complex molecules from smaller molecules are called anabolic. They create living matter. Catabolic processes, in turn, support anabolic processes by creating energy from breaking down larger molecules. Living systems, such as cells, balance the demand for growth and energy production.

An organism takes in resources and products that other organisms consider waste, transforms these flows to sustain life, and expels "waste and by-products" that other organisms consider resources or products. These dynamic exchanges of life-sustaining flows also occur at different scales of natural systems. Here is a simplified example that illustrates these principles.

When we breathe in air, oxygen is captured by blood in our lungs and transported via blood cells to the tissues throughout our bodies. Once the oxygen is metabolized within the cells, carbon dioxide is expelled and carried back by blood cells to the lungs where it is exhaled out of our bodies. Trees and other plants take in the carbon dioxide to use in photosynthesis. One of the byproducts of photosynthesis is oxygen. Thus, the cycle of exchange continues—waste from one organism becomes a life-sustaining resource for another. Similar processes and relationships exist within and among organizations participating in business ecosystems. These relationships have been shown to cut costs, reduce and prevent pollution, and increase business and employment opportunities (see Chapter 5).

For example, a typical electric power company and its community of suppliers, "metabolize" many resources, such as coal, water, natural gas, copper, iron, petroleum, and timber, to make, distribute, and sell electricity to their customers. A comprehensive metabolic pathway within this business ecosystem includes all resources used, and the products and by-products produced in the production, distribution, and selling of electricity. This includes, for example, the mining, extraction, and transportation of coal, and the manufacturing and installation of copper wires, transformers, and utility poles. Other individual organisms along this metabolic pathway, such as a coal mining company, metabolize resources to produce products, services, and by-products that are passed along the "food web," some of which are dispersed as "waste." In general, a company's metabolism involves a dynamic combination of both catabolic and anabolic pathways. As already illustrated in Chapter 2, this life-cycle thinking has significant implications for sustainable development.

In the case of the Asnæs Power Station in Kalundborg, Denmark, discussed in Chapters 1 and 5, many of the power plant's metabolic by-products are exchanged with other companies, such as the Novo Nordisk pharmaceutical plant, the GYPROC wallboard manufacturing plants, and the Kemira chemical plant. Exchanged by-products from just the power plant's metabolism include sulphuric acid, heated water, steam, fly ash, and gypsum. Metabolic by-products from surrounding "organisms," in turn, flow back to the power plant. For instance, cleaner-burning refinery gas from the STATOIL refinery is metabolized by the Asnæs plant, instead of 30,000 tons of coal, thus reducing air emissions and operation and maintenance costs.

However, the Kalundborg community is only a small portion of a broader business ecosystem, or "food web," that supports it. Other subsystems include coal mining, extraction, and transport; oil and gas exploration and development; and manufacturing of equipment and technologies used by Kalundborg companies. Ultimately, this "food web" is supported by resources derived from a "natural economy." And, like natural systems, metabolism occurs at different scales—company, business community, and business ecosystem.

It is important to note that organisms satisfy their metabolism first and then "sell" residual products and flows to other organisms. Similarly, a life-sustaining organization can sustain and improve its own metabolism by providing internal products and services. Examples include employee training, health benefits, on-site child care, equipment redesign and modernization, performance/production incentives, process design and efficiency improvements, and creativity workshops and retreats. Smaller firms may improve their metabolism by participating in collaborative networks, cooperatives, and associations that provide similar products and services.

Business ecology combines common business sense with insights from natural systems. As shown in Figure 3.1, your company's internal economy—its metabolism—has three vital, interrelated processes:

1. Values-based organization and development
2. Value creation
3. Cyclical flows

Any business, no matter what type or size, requires these three core processes to be healthy or viable. In viewing Figure 3.1, think of these processes as interdependent, fluid, self-organizing systems that dynamically interact within your organization and collectively exchange flows with the external environment. The following sections describe each of these processes and how they relate to each other, to help you see your company's metabolism. While the focus here and throughout the book is often on for-profit organizations, these discussions apply to all organizations, including nonprofits, government institutions, and communities.

VALUES-BASED ORGANIZATION AND DEVELOPMENT

In business ecology, values-based organization and development includes: the business plan, company vision and values, policies and guidelines, bylaws, governing bodies (e.g., Board of Directors), human resources, project management, training, and all the administrative activities, such as accounting, payroll, and tax payments that keep a company in legal compliance and financially solvent. These essential elements give structure and shape to your organization, define its purpose, and help maintain its viability.

The darkly-shaded object labeled "values-based organization and development" in Figure 3.1 represents the nucleus of your organization. **It is** here that we find the natural organizational blueprint—social DNA—for your company's development. Joseph Levine and Kenneth Miller briefly explain the astonishing role of DNA in *Biology: Discovering Life*:

> In a sense, every living cell is a storehouse of the very information about the nature of life that biologists have always sought. . . . Each DNA (deoxyribonucleic acid) molecule is a double string of billions of atoms coiled up tightly into a beautiful and elegant spiral chain. Inscribed on the chain is a wondrous code, a linear sequence of four simple molecules that carries the information necessary to build and operate a fly, an oak tree, or human being. It is a code that was well-conceived along with life itself nearly three and a half billion years ago, and a code that we are just beginning to decipher.

Why bother with an explanation of how DNA works as the organizational blueprint for living systems? Because your organization is a living system. DNA directs the synthesis of proteins (or enzymes), which in turn catalyze reactions to form other biological molecules, which in turn comprise the living matter that constitutes the cell. Cells, as we know, are the basic units of life. DNA helps control the day-to-day activities of the cell as well as the development and reproduction of simple and complex organisms. Organisms can be single-celled, like bacteria, or as complex as a human being. In terms of this discussion, organisms are organizations— your business or nonprofit for example.

A cell, the most basic unit of life, is both centrally and decentrally organized. The nucleus, containing a pool of genetic material, dynamically interacts with the internal environment within the cell, or the cytoplasm. DNA and other genetic messengers move back and forth from the nucleus into the cytoplasm, helping direct the synthesis of living material, life functions, and reproduction. Proteins called enzymes create the conditions for combining and breaking apart materials, depending on the need (e.g., building biological material or releasing energy to support living functions). **The cell's management is not top down. Its nucleus is fully integrated with other parts of the cell.**

Without being molecular biologists with Ph.D.s, how can we take lessons from life and living systems and apply them to our organizations? **First, an organization's social DNA is its values.** They affect how organizations behave, develop, and interact in their environment, particularly in times of turbulence and uncertainty. People are like the cells in an organism, the basic units of life. People, of course, have their own values that make them unique. We are speaking of the shared values that define an organization's identity and purpose. The degree to which shared values are understood, accepted, and practiced defines the level of community within an organization. Like DNA and cells, community-based organizations are not threatened by change, but evolve to meet challenges and realize new opportunities. Later chapters explore these concepts more deeply.

Natural systems, from a single cell to the entire biosphere, create conditions for life; they do not manage or control it like a machine. Team-building, open-book management, and learning organizations are examples of recent organizational trends that, to some extent, mirror natural systems. These trends, for example, stimulate creativity, foster collaboration, create a more open flow of information, and produce higher levels of organizational knowledge and competence.

Nature's way of organizing creates a dynamic dance between chaos and order. Creating the right mix of both can make innovation and success not only possible, but fun. Here's what Xerox Business Services' Chris Turner has to say about their work environment in her interview with *Fast Company* magazine in 1996:

> What we're learning is a different way of doing business, a different way of engaging people. How do you measure it? Every piece of common sense tells you that if you have a place where people come to work and have fun and find meaning and feel as if they make a difference, that's a place where the profits will be better.

Value Creation

Value creation is a cyclical, incremental process that makes products or services useful and marketable within and outside an organization. Life-sustaining flows—money, energy, products and services, information and ideas, people and other organisms, air, food, water, and materials—contribute to the value-creation process. Like value creation in living systems, organizations use anabolic and catabolic processes to create value. For instance, cells synthesize proteins into living matter, through anabolic processes guided by genetic codes, by using energy produced by catabolic reactions in the mitochondria of a cell—its power plants. Similarly, a factory process combines raw materials, products and services, design information (social DNA), and energy produced on site or by a power plant, to produce and process viable, value-added products and services.

Value creation is what sustains an organization and its stakeholders economically. It is the "billable" work and the work that pays your wages, dividends, and bills. Related to value creation, business ecology:

1. Uses full-cost accounting to determine *net value* created by a product or service. This is the difference between price, a reflection of social benefits received from a product or service, and its total life-cycle costs.
2. Distinguishes between *actual* value creation, where quality of life genuinely improves, and *apparent* value creation, the value improvement that we believe or expect to receive that may or may not be realized.
3. Seeks to optimize a broader set of wealth-creating cyclical flows within and outside an organization.
4. Recognizes that value created by an organization extends beyond the customer and shareholder to a broader community of stakeholders.

Business ecology considers the *net value* of products and services. Net value is the difference between market price, a surrogate measure of the actual value received from a product or service, and its total life-cycle costs. The key point here is the difference between traditional accounting and ecological accounting. Ecological accounting asks: What are the total costs and benefits along a value-creation cycle? Externalities such as pollution, worker displacement, ecological destruction, waste disposal costs, and reduced quality of life are often only partially reflected in the price. Remaining costs are generally borne by the public through taxes, regulations, infrastructure, and the environment (e.g., reduced or degraded habitat). Business leaders such as Monsanto's Robert Shapiro and Interface's Ray Anderson believe more accurate, ecological accounting, where externalities are brought into a company's internal decision process, will help stimulate sustainable, ecologically efficient business behavior. Such integrated accounting of business activities creates an environment for ecologically efficient competition.

Business ecology distinguishes between *actual* and *apparent* value creation. Actual value creation improves quality of life for customers and other stakeholders. Their quality of life, for instance, is improved by products and services; by enduring, good-paying jobs; by a stronger, more vibrant economy; and by other recognizable forms of value created by an organization. In contrast, apparent value creation is the quality of life improvement that we believe or expect to receive from a product, service, or an organization, which may be greater, lesser than, or equal to actual value creation. Here are two examples that illustrate why this distinction is important.

In our consumer society, apparent value creation is often cultivated to ensure that the consumer believes certain benefits will be gained by a new product or service, whether or not he or she actually experiences the expected quality of life improvement. For example, prior to the winter holiday shopping season, a blizzard of commercial advertisements creates wants, disguised as needs, among children. Demand for a new toy may be created by linking the product with popular television programs and celebrities. The child may experience a short period of satisfaction from the toy, but soon it is forgotten or displaced by the next popular toy. In this case, the apparent value creation—the expectation—is greater than the actual value. The result: apparent value creates overconsumption in a world of limited resources and growing poverty, while the quality of life in consumer-based societies actually erodes from the misallocation of time, income, life energy, and other life-sustaining flows.

Value creation induces the flow of money, energy, and other life-sustaining flows toward an organization. Wasteful, inequitable, and destructive consequences can result when decisions are based on apparent rather than actual value creation. Here's another example: Decisions to downsize a highly-skilled workforce, close plants, and cut research and development spending during periods of high corporate earnings, while driving up shareholder earnings and the apparent value of a company in the short term, can erode the long-term, actual value-creating potential of a company, especially when the entire stakeholder community is considered. Such short-term decisions often create permanent, irreversible economic problems that can ripple out into the stakeholder community, affecting communities where plants are located, suppliers, and government investments in infrastructure. Further, damaged stakeholder relations may eventually lead to lower corporate earnings. For example, in his book, *The Living Company*, Arie de Geus describes this conflict between economic profits and longevity:

> [Senior managers] are simply trying to do their best against the background of widespread misapprehension of corporate purpose. They are struggling with a terrible dilemma. When shareholders and outside regulatory bodies ask about the senior mangers' results, they do not ask about efforts to improve community. They do not inquire about the company's prospects for a long and prosperous existence. They ask, "What is your return on capital employed?" "Aren't you overcapitalized?" and "What is your productivity?"

De Geus cites a well known example—Exxon:

> Exxon let 15,000 people go in 1986, in the wake of the oil price col-
> lapse. They concentrated power in a narrow chain of command and
> took away one side of their organizational matrix structure. In the
> process, they considerably reduced their managerial capacity. A year
> later, the Valdez oil spill incident took place. It took them 48 hours to
> react. That 48 hours has so far cost them $3 billion in cleanup costs,
> bad publicity, and legal fees. And the ticker is still counting.

Stakeholders of an organization, such as customers, employees, shareholders, and suppliers, contribute and/or receive value along the cycle. For
example, customers experience recognizable benefits from products and
services that actually meet or exceed their needs, desires, and expectations;
shareholders and creditors receive returns on investments; employees are
compensated with salaries, health benefits, and equity shares; federal, state,
and local governments receive tax revenues both from the company and its
employees; and nonprofit groups receive tax-deductible donations. These
value flows, in turn, support value creation in other organizations.

The life-cycle of value creation involves the formative environment of
an idea or innovation; planning, research, and development; purchasing,
acquisition, and outsourcing; design and testing; production and processing; marketing and sales; and the return flow of cash, products, and services, and other life-sustaining flows.

Cyclical Flows

Cyclical life-sustaining flows such as energy, information and ideas, materials, products and services, water, air, people and other organisms, food,
and money contribute to the value creation process within and outside an
organization. Figure 3.1 shows each of these flow systems moving into and
out of the business organism. Chapter 4 describes how each of these cyclical
flows is important to every organization, and how the relative importance
of different flows—such as food in the restaurant business or information
and ideas in the publishing industry—define and distinguish metabolism,
niche, and habitat of different sectors and industries.

Business ecology provides a framework for improving your organization's ecological efficiency and viability. It also provides ecological knowledge of other organisms that might share cyclical flows—an essential step
in developing healthy business ecosystems (see Chapter 5). In the emerging
closed-loop, ecological economy, value creation occurs throughout prod-

uct/service life cycles, and becomes increasingly ecologically efficient. For example, less energy, materials, and other life-sustaining flows are consumed in product manufacturing and processing. In the return flows from customers, retailers, and other stakeholders, "spent" products and by-products are recycled. Thus, companies are moving toward cyclical, resource-efficient behavior and away from linear, wasteful practices of the industrial era. The recent development of "recycling," "demanufacturing," and "remanufacturing" industries represent an important step toward the ecological economy.

Closing the loop can often be profitable within a company. William S. Bailey, Jr., Material Management Manager with Seminole Electric, describes the benefits of an internal resource recovery plan:

1. Materials Management makes an income contribution to the company.
2. This process lowers the total material cost.
3. Assists in maintaining optimum inventory levels.
4. It maximizes materials movement.
5. It provides a focus on corporate resource recovery.
6. Satisfies social and environmental goals of recycling and not wasting resources.

Bailey describes investment recovery as a logical extension of the purchasing-materials cycle within organizations. Bailey claims that, while purchasing and materials management professionals are familiar with investment recovery, many companies do not have effective programs. As a result, these companies often do not capture benefits from investment recovery such as lower costs, additional income, and more efficient use of materials, time, and space. Here he outlines why and how investment recovery works within organizations:

Investment Recovery [IR] provides two main services within the organization. First, it provides a central function within the company for employees to come to when something needs to be disposed. Second, a managed outlet for resources or materials that either finds a market for it and recovers funds or disposes of through other methods:

1. Re-use at another location within the company
2. Sell directly to another company or dealer
3. Return to supplier for credit
4. Trade in
5. Recondition/rebuild

6. Scrap, dismantle, or destroy
7. Classify as waste and dispose of accordingly
8. Donations
9. Material loans to other organizations

NICHE: WHAT YOUR COMPANY DOES
FOR A LIVING

Finding your market "niche" is essential to any healthy business. Already well accepted and widely used by the business community, the concept of niche has even more significance in the context of business ecology. How is niche defined? *Webster's Ninth New Collegiate Dictionary* (1984), provides powerful insights:

(1) a: a recess in a wall esp. for a statue; b: something that resembles a niche; (2) a: a place, employment, or activity for which a person or thing is best fitted; b: a habitat supplying the factors necessary for the existence of an organism or species; c: the ecological role of an organism in a community esp. in regard to food consumption.

The book *Biology: Discovering Life* provides a simple, yet highly useful definition of a niche: "how an organism makes its living." The authors Joseph Levine and Kenneth Miller describe three important factors that determine a niche:

1. There are a number of vital flows and conditions that sustain an organism and allow it to reproduce. Examples of vital flows include energy, materials, information, nutrients, and water. Example conditions include temperature, humidity, salinity, pH, grain size of soil, and similar variables.
2. Organisms interact with other organisms. A number of biological relationships affect an organism's survival and success. These include competition, predation, parasitism, organisms that provide shelter, and organisms that collaborate.
3. Niche is also determined by an organism's behavior. This includes when, where, and upon what it feeds; an organism's social system, if any; and its behavioral patterns with other organisms.

Business ecology defines niche as how your business "makes a living" within its environment. It is worth taking a moment to consider the biological relationships, such as competition, parasitism, and predation, that exist between your company and its various stakeholders. Suddenly, expressions like "It is a jungle out there," or "I'm swimming with sharks!" take on real significance. There really is a "business environment"! Your company has "biological relationships." It has vital relationships with the economy, the community, and the environment. This includes the stakeholder community—all those individuals and organizations that affect, and are affected by, the existence of your company. These can include: customers, shareholders, creditors, suppliers, manufacturers, distributors, utilities, retailers, citizen and community groups, various levels of government, contractors, accountants, lawyers, competitors, and future generations. "Making a living" in this context includes how your company communicates and interacts with its stakeholders.

By understanding the web of these stakeholder relationships you can position your company favorably within its environment. Looking again at Figure 3.1, it is clear that several factors affect how your company "makes a living" (niche) and "where it lives" (habitat—see next section). Life-sustaining flows include products and services, money, information and ideas, people and other organisms, air, energy, food, materials, and water. Depending on the type of business, these flows have varying degrees of importance. The timing of these flows, as well as space requirements of a business, also affect its niche and habitat.

As always, an organization's core purpose and values, its social DNA, define its identity and behavior and generally act as fixed references. However, as is the case in nature, organizations must be adaptive and open to changes in their environment. Organisms respond to changes with short- and long-term strategies. Short-term strategies include changes in behavior (e.g., climatic changes alter habitat and feeding habits of migratory birds and butterflies) and appearance (e.g., the arctic fox changes its color to blend with seasonal changes in its surroundings). Longer-term strategies infer deeper, often irreversible changes such as different genetic expressions within species (e.g., different types of foxes) from mixing of DNA and the introduction of a brand new species (e.g., humans). Here are examples that show how organizations develop short- and long-term adaptive strategies to respond to changes in their environment.

For instance, a franchise such as McDonald's may change its external appearance to historic in a colonial city with strict building requirements or modern in a bustling, downtown metropolitan area. Its menu may also vary to reflect changes in season, low-fat diets, and cultural preferences. But even with these variations, McDonald's restaurants generally hold to their core values—convenience, superior value, and excellent operations—they are clean, offer fast food of consistent quality, and provide responsive, customer-oriented service.

But adaptive strategies can occur on a deeper level as well. Companies such as Monsanto and Interface are adapting their core values to embrace sustainability. For example, they have enlisted The Natural Step (TNS) to help their companies evolve along this path (see Chapter 7). Interface now publishes a Corporate Sustainability Report and holds itself accountable for following a path to sustainability that includes seven measurable steps:

1. **Eliminate waste** The first step to sustainability, QUEST is Interface's campaign to eliminate the concept of waste, not just incrementally reduce it.
2. **Benign emissions** Prioritized focus on the elimination of molecular waste emitted to natural systems that have negative or toxic effects.
3. **Renewable energy** Reducing the energy demands of Interface processes while substituting non-renewable sources with sustainable ones.
4. **Closing the loop** Redesigning Interface processes and products into cyclical material flows.
5. **Resource efficient transportation** Exploring methods to reduce the transportation of molecules (products and people) in favor of moving information. This includes plant location, logistics, information technology, video conferencing, e-mail, and telecommuting.
6. **Sensitivity hookup** Creating a community within and around Interface that understands the functioning of natural systems and our impact on them.
7. **Redesign commerce** Redefine commerce to focus on the delivery of service and value instead of the delivery of material. Engage external organizations to create policies and market incentives encouraging sustainable practices.

On a global scale, economic, social, and environmental challenges are creating a values shift in business. The limits to global resources and environmental issues such as habitat destruction and climate change are changing the rules, the social DNA, of commerce. Like organisms responding to a changing environment, businesses are evolving from extractive, wasteful practices to more closed-loop, resource-efficient behavior. Many companies such as Interface, IKEA, Mitsubishi, and Monsanto see sustainability as a new core value and a strategic business opportunity. Just a decade or two ago, sustainability and environmental issues were perceived as peripheral to their businesses. Tachi Kiuchi, managing director of Mitsubishi Electric Corporation, describes his company's short- and long-term, adaptive strategies to *Business and Environment* (October 1997):

> We want to create corporate feedback so we can know our environmental costs and benefits better than any other company" . . . [Kiuchi] said that starting in 1998, Misubishi will have indicators for pollution intensity, resource productivity, and quality of life. He talked about his business having a purpose and meaning beyond profit: "to fully develop the human ecosystem, so we can consume less, and be more.

A Comparison of Hypothetical Niche Enterprises

These examples are based on actual facts and conditions described in more detail in the Appendix. Three hypothetical examples—a franchise bagel sandwich shop, a community-based grocery store, and an ecologically efficient bakery/brewpub/restaurant—show how, in the same habitat, different niches are filled by similar and sometimes competing businesses. Table 3.1 compares these hypothetical niche enterprises in Inner West Street, Annapolis, Maryland. These examples illustrate a number of similarities between the business "niche" and the ecological "niche," and illustrate a number of important relationships that can develop within and among businesses, other organizations, the community, and the environment. Businesses develop with respect to the web of sustaining flows and relations, including: organizational form, source of start-up funding, source of core values/purpose (social DNA), type of business, products and services, target customers/key relationships, source of employees, space and habitat characteristics, supply web, and resource/product loops and flows.

First, like animals and plants in an ecosystem, businesses often position themselves favorably with respect to vital resource flows, space, time, and other organisms. In scenario A, the bagel bakery/sandwich shop is very dependent on customers from the medical center and professional offices, and on equipment and ingredients from the national franchise. This dependency, however, is balanced by well-paid, repeat customers and the "know-how," training, experience, and reliability of a franchise. Scenario B shows how strong, self-reinforcing community ties can lead to win-win outcomes for both business and community. For instance, a community with employed citizens and shared vision is more cohesive and healthy, as well as less prone to crime. Such neighborhoods, in turn, often attract more

Table 3.1 A Comparison of Hypothetical Niche Enterprises in an Urban Neighborhood

Niche Characteristics	Scenario A	Scenario B	Scenario C
Organizational Form	Franchise/Sole Proprietorship	Community-Based Enterprise/Limited Liability Partnership	Limited Liability Partnership, Assoc. with Brewpub/ Restaurant
Source of Start-up Funding	National Bank	Government Grants, Local Banks	Company-Financed (Brewpub/ Restaurant)
Source of Core Values and Purpose (Social DNA)	Franchise/Owner	Community	Brepub/Restaurant Owner
Type of Food Business	Bagel Bakery/Sandwich Shop	Grocery Store/ Bakery	Bakery, Attached to Brewpub/ Restaurant
Products & Services	Bagels, Bagel Sandwiches, Desserts, Coffee, Herbal Teas	Meat, Fish, Produce, Dairy, Household Goods, Baked Goods, Pharmacy, Video Rentals	Breads, Rolls, Breadsticks, & Pretzels
Target Customers/ Key Relations	Professionals/Medical Center, Tourists; Depends on Commuters & Franchise	Residents of West Street Neighborhood & Offices; Depends on County/ Neighborhood	Singles, Professionals, Tourists, Employees; Depends on Brewpub/ Restaurant, Neighborhood
Source of Employees	High School/College Students	Clay Street Neighborhood Adults & Teenagers	Brewpub/Restaurant Employees, Students
Space & Habitat Characteristics	1400 sq. ft. Parking, Convenience	10,000 sq. ft. Pedestrian, Selection & Convenience, Parking, Delivery	800 sq. ft. Parking & Pedestrian Entertaining
Supply Web	International/National	Regional/Local	Regional/Local, "Closed Loop"
Business Hours	M-F, 6:30 am – 3:30 pm Sa, 8 am – 3:30 pm Su, Closed	M-Sa, 9 am – 10 pm Su, 10 am – 5 pm	T-F, 7:30 am – 8 pm M, Closed Sa, 7:30 – 5 pm Su, 7:30 – 4 pm

Table 3.1 illustrates a number of similarities between business "niche" and ecological "niche." Much like animals and plants in an ecosystem, businesses often position themselves favorably with respect to vital resource flows, space, time, and other organisms. Each of the three food businesses—a franchise bagel sandwich shop, a community-based grocery store, an ecologically efficient bakery/brewpub/restaurant—have developed different strategies (niches) for competing in the same resource-limited habitat. Their niches are defined by a number of factors, including organizational form, source of start-up funding, source of core values/ purpose (social DNA), type of business, products and services, target customers/key relationships, source of employees, space and habitat characteristics, supply web, and business hours.

businesses. Scenario C illustrates how looking more systemically at a business, or two adjacent businesses, can lead to more efficient, closed-loop resource exchanges. In this case, more efficient use of energy, employees, food, and space combine to yield quality products at lower costs. Finally, this example also shows how an existing business, such as the restaurant/brewpub, can help anchor and sustain a "sprouting" bakery business, much like "pioneering species" help establish conditions for subsequent species introduction and the development of mature, diverse ecosystems.

Another aspect of niche is how organisms try to share their habitat and compete for available resources. This includes preferring different kinds of food, obtaining food and shelter in different places, and being active at different times of day. Similar relationships often develop among businesses. For example, the three scenarios—all businesses selling food in the same part of town—have subtle differences in locations, as well as different target customers, business hours, and products and services. Their key relationships with customers and other stakeholders and their social DNA are different—one is a franchise, the second is a community-developed enterprise, and the third is an outgrowth from an existing individually owned business—which is why these enterprises can coexist in shared markets.

HABITAT: WHERE YOUR COMPANY LIVES

There are many documented cases where businesses outperform competitors by simply being in the right location. This is the power of habitat. A business's habitat is determined by the same life-sustaining flows that affect niche: products and services, money, information and ideas, people and other organisms, air, energy, materials, food, and water. Here are two examples that illustrate business habitat: (1) a shopping mall, the forces that dominate its centralized, car-dependent design, and how it fits into a larger metropolitan development system; and (2) the dispersed, customer-oriented habitat of Xerox Business Services, which illustrates how forces, such as technology and work patterns, are creating new business habitats.

The Shopping Mall Habitat

Shopping malls are business ecosystems that commercial developers often locate at the intersection of highways to maximize automobile access by consumers. In some instances, malls also are located with convenient access to subways, trains, and even neighborhoods. Overwhelmingly, malls are designed for one-stop shopping; the shoppers can arrive in cars and are a captive audience for the many vendors within the mall. Tables and benches encourage socializing in conjunction with spending money on shopping, food, movies, video arcades, and other forms of entertainment. While malls are becoming today's village centers, providing a place to socialize particularly for teens and seniors, make no mistake: the primary design goals—and values—are easy automobile access and profits from sales for the individual retailers and the developers who own the mall. In addition, large parking areas surrounding malls are designed for the weekend and holiday shopping periods. While packed during these periods, they often remain underused during the remainder of the year; this is an inefficient use of land and resources.

The collaborative model at work among mall business "organisms," however, deserves a closer look. In a mall, businesses and vendors lease space and receive the benefits of collaborative marketing, co-location, one-stop shopping, customer spillover, shared parking, an enclosed, year-round shopping environment, and other advantages. There is great potential for applying this model to clusters of small businesses in towns and neighborhoods, where a strong support network, such as a local Chamber of Commerce, does not yet exist or is not well established. For instance, similar businesses, such as art galleries, can work together to market themselves with special events and publicity. Or a neighborhood may promote its shopping district with a brochure or flyer on their diverse offerings—or even a street festival.

At a mall, anchor organisms, such as department stores and movie theater complexes, create the main draw for customers. Smaller retail, specialty stores, and open-air vendors position themselves in the corridors connecting the larger mall organisms. Restaurants, coffee shops, and bakeries cluster in food courts to make mealtime convenient so consumers can continue shopping. Automatic teller machines are strategically placed to provide ready access to cash and stimulate consumer spending. Special events and promotions, such as "Outlet Mall Week," fashion shows, and even "water parks," create additional attractions to bring in customers. In some instances, parking garages are provided to offer protection from the weather and reduce the consumers' walking distance from their cars to the mall entrance.

How do malls fit in the broader context of urban metropolitan development? As urban sprawl expands outward from older, decaying inner cities, shopping malls are key outposts of the spreading suburban environment. New malls are strategically positioned in the outermost growth ring of suburbs, at the intersection of major transportation arteries. This is the mall's habitat within the regional business ecosystem. In contrast, ghost malls and abandoned or declining downtown areas often dominate areas of previous growth. Providing regional infrastructure, such as schools, roads, rail, and sewage, is costly and difficult in these outermost development areas, while existing infrastructure in "old growth" areas is underused. Vast areas of land and environmental resources are also consumed with urban sprawl. For these and other reasons, sprawl development undermines our long-term economic, social, and environmental sustainability.

Shopping malls, while not sustainable, do serve as useful models for understanding business ecosystems and habitat. Examples of how to develop healthy malls and business ecosystems, including redevelopment of "old growth" economic areas, are included in Chapter 5.

The Dispersed Business Habitat

While a shopping mall represents a centralized, collaborative habitat that attracts customers to businesses, many businesses have dispersed habitats. Businesses with dispersed habitats come to the customer, or even share the same space with the customer. Technologies such as the Internet, cellular phones, groupware, and notebook computers are changing where a company lives and where work is done, and dramatically expanding the marketing reach of even the smallest of companies. When these technologies are coupled with a creative atmosphere, a company can exist in a "dispersed habitat" that is close to or even within their customers' environment. Xerox Business Services (XBS), with headquarters in Rochester, New

York, is such a company. Chris Turner, a fifteen-year XBS veteran charged with developing XBS's change strategy, was interviewed by the founding editor of *Fast Company*, Alan M. Webber (October/November 1996). The following is derived from that interview.

XBS is a fast-growing offspring of Xerox, the copying and document management company based in Webster, New York. More than 80 percent of XBS's employees work at customer sites rather than XBS facilities. That is, the XBS employees' offices are in the same buildings as their customers. In a very real sense, XBS serves as a neural network that links onsite XBSers, customers, and a highly dispersed, 15,000-member XBS learning community. XBS is remarkably successful—a billion-dollar company with a 40 percent annual growth.

What makes XBS so successful? How does its dispersed workforce of 15,000 remain cohesive and focused? Developing an energetic, creative learning environment with a shared vision; building trust, self-knowledge, and strong relationships; and seeing organizations as natural systems are key features of their success. XBS's learning environment is relationship- and values-oriented; it encourages team-building, personal development, and rewards collaborative results. Committed and innovative employees cultivate satisfied customers. These customers include Intel, Microsoft, General Electric, Motorola, Lufthansa, TRW, Dow Chemical, and Texas Instruments, and a host of smaller companies. Chris Turner, whose job is to stimulate learning and positive change within XBS, had this to say in her interview with *Fast Company*:

> Reengineering is dead. It was the last gasp of the old rigid command- and-control corporate model. We have the sense that many people are beginning to look in a new direction. We know the outlines of it, but the precise shape it will take is not exactly clear. We've learned that you can't push change through an organization, because the harder you push, the more resistance you get. That's why we've turned to the concept of natural systems. We've also learned that work life should be more like home life. People work better when they don't pretend. And we've learned that you have to trust people to develop their own solutions in the workplace. When they own them, they make them suc- ceed.

The high energy and creativity of the XBS learning community is con- tagious, particularly since it achieves business results. In the context of busi- ness ecology, XBS's core values such as learning, trust, creativity and decentralization, are its social DNA. They drive the living organism that

XBS is. They affect how XBS does business, and how it grows and develops. By sharing the habitats of its customers, XBS does not guess what its customers want; it creates value by asking them directly, anticipating problems, developing solutions, and tapping the creative energy of 15,000 XBS employees. XBS serves more than 4,000 customers in thirty-six countries, and will likely stimulate change by sharing its core purpose and values—its social DNA—with its customers and other stakeholders.

APPLYING BUSINESS ECOLOGY TO AQUACULTURE

Niche, habitat, living organisms—to illustrate the lens and framework of business ecology we will apply it to a specific industry: aquaculture. We will view the aquaculture business organism in its environment to discover how social DNA, cyclical flows, and stakeholder relations affect its metabolism, niche, and habitat (see Figure 3.1). We will also compare different forms of aquaculture, to gain insights into how social DNA—an organization's core purpose and values—shapes and develops it, particularly in the context of sustainable enterprise and value creation.

Aquaculture, the cultivation and harvesting of aquatic animals and plants, holds considerable promise in the next economy. Worldwide fisheries are severely overharvested, with most at or near collapse. The demand for seafood continues to rise with the population. Many species, such as striped bass, mussels, crayfish, shrimp, sturgeon, and oysters, are being cultivated to replace the depleted natural stocks. Aquaculture has a strategic role in both helping to restore natural resources and providing food and other products and services. Examples of nonfood products and services include: pharmaceuticals, biotechnology, pollution control, ecological restoration, energy, fertilizers, bait, and ornamental plants and animals.

Figure 3.2 shows aquaculture through the lens of business ecology. Cyclical flows (such as energy, food, money, and water) and stakeholders (such as suppliers, energy providers, and customers) contribute to and receive from the value-creation process. Commercially viable, value-added products and services, in turn, are marketed, sold, and distributed to customers. The elongated shape with embedded elliptical bodies represents the "value-creation" system within the aquaculture organism. Vital flows are dynamically exchanged among the aquaculture organism, its stakeholders, and its environment through its membrane.

Value creation, efficient use of cyclical flows, and the building of healthy stakeholder relations, as these examples will show, are vital to a

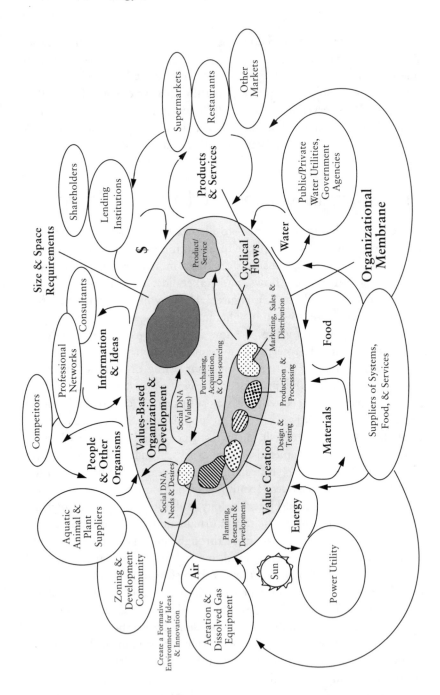

Figure 3.2   The Aquaculture Organism and Its Environment

© 1998 Business Ecology Associates

Table 3.2 Profile of the Aquaculture Organism

Habitat/Niche	*Key Stakeholders/Considerations*
climate and environmental considerations	tradeoffs between distance to market, energy, and labor costs
types of aquaculture products and services	customer needs/desires, technical/training requirements
biological requirements of aquatic animals/plants	water quality, physical setting, difficulty raising/breeding
access to transportation	rail, water, highway, air, pedestrian, embodied energy costs
location, space, and characteristics of markets	tradeoffs between transport/processing/land costs and market access
Flows That Create Value	
financial	banks, government agencies, perceived investment risk
information and ideas	professional organizations, government agencies, universities, consultants
energy	proximity to power plants/residual energy sources, solar
food	manufactured versus grown, cost, quality, proximity to food sources, climate, processing and management
water/air	salinity, temperature, pH, dissolved solids and gases, nutrients, water availability, regulatory environment
people and other organisms	labor costs, availability of well-trained technical and business personnel
products and services/materials	full-service providers: pumps, tanks, monitoring, filters, feeders, aerators, greenhouse structures

© 1998 Business Ecology Associates

competitive, sustainable aquaculture enterprise. Table 3.2 provides a profile of the aquaculture organism. Important factors affecting its metabolism, niche, and habitat include: climate and environmental considerations; access to transportation, space, energy, food, water, and nutrients; location and characteristics of markets; biological requirements of aquatic animals and plants; labor costs; and availability of well-trained technical and business personnel.

Climate and environmental considerations determine both the nature of the aquaculture enterprise and its territory. Tropical and subtropical regions such as Ecuador, Southeast Asia, and the Philippines, have experienced high growth in aquaculture, particularly with high-value products such as shrimp. The year-round production is sustained by the ample solar radiation and the availability of cheap labor. Funding from groups like the World Bank has also stimulated aquaculture growth in these regions.

However, in the Southern Hemisphere, short-term economic opportunity is often given priority over long-term, sustainable management of natural resources. This is particularly true where enforcement of environmental laws is weak or nonexistent. These practices, which have been described as "rape and run" development, have given aquaculture a bad reputation. For example, environmentalists often link aquaculture with the destruction of mangrove swamps. For this and other reasons, many financial groups, such as the World Bank, are becoming increasingly sensitized to the consequences of their funding programs and are looking for ways to encourage more sustainable development of aquaculture.

Many within the aquaculture community are recognizing the need for sustainable development of their industry. As shown in Figure 3.2, professional organizations are often a good source of information and ideas and, when well organized, can shape an industry's development. For example, the World Aquaculture Society featured "Sustainability" in their June 1996 journal, *World Aquaculture*. In February 1997, they held a conference in Seattle, Washington, "Linking Science with Sustainable Industry Development." Publications and events such as these can propagate change throughout this industry by creating a collective knowledge network among stakeholders, and help market aquaculture as a sustainable industry.

Access to energy, transportation, space, food, water, and nutrients are key factors affecting the aquaculture business environment. Excess energy from power plants and other residual energy sources, such as refrigeration plants, can provide strategic opportunities for aquaculture in colder climates. In the Kalundborg, Denmark, example introduced in Chapter 1, the fish farm uses residual heat from a power plant. It produces enough fish to export abroad.

The State of Maryland is an example of a cold-weather aquaculture habitat in the U.S. While Maryland has a colder climate and higher labor costs than tropical habitats, it is well-situated relative to transportation and upscale customers. For example, striped bass harvested in Maryland can reach 40 percent of the U.S. population within one day if delivered by truck. Energy and food processing plants in the region offer significant sources of cheap heat and nutrients to offset the colder climate. Available rail and water transport also support the industry's development. In addition, Maryland is blessed with a world-class fisheries resource, the Chesapeake Bay, and rich, productive agricultural lands flanking its eastern and western shorelines. Finally, Maryland is one of the few states which funds a full time position (Department of Agriculture) to support the development, growth, and promotion of aquaculture.

The cost of land or space can have obvious impacts on the viability of an aquaculture enterprise. While many operations occupy sites exclusively in low-cost farm areas, there are many examples of aquaculture co-existing with other economic and social activities. For example, Chinese agro-ecosystems frequently integrate both land and aquatic production. Shrimp, rice, and hogs, for instance, may be raised on the same parcel to encourage efficient recycling of resources—such as pig manure to replenish the land and rice detritus to feed shrimp in flooded areas. With recirculation systems, aquaculture is developing in urban and town centers closer to the market. Abandoned urban property and rooftops are becoming important sites for urban aquaculture-hydroponic-composting systems, particularly when markets are nearby and residual heat and other resources are available. Chapters 4 and 5 provide examples of such integrated systems, including the Dan'l Webster Inn in Cape Cod, Massachusetts, and Bioshelters, Inc. in Amherst, Massachusetts. "Freshness" and "pollutant and chemical-free production" are key marketing advantages for such companies, particularly with health-conscious and environmentally aware customers.

Many aquaculture companies use pelletized feed from suppliers, because it is easier to automate and control feeding. Similarly, they may use fertilizers to grow aquatic plants. As shown in Figure 3.2, some companies provide a complete line of food, systems, and services to support aquaculture producers.

Growing feed is an inexpensive way to supplement manufactured feed. Processing food scraps derived from restaurants (such as shrimp meal) and food processors can boost the protein content inexpensively. Effluent from aquaculture and water reclamation facilities provides an excellent growing medium for duckweed and other plants that can be harvested as potential food supplements. Proximity to low-cost, high-quality food and nutrient sources can dramatically improve production and reduce operating costs.

Depending on the method and type of production, the quantity and quality of the water directly affect costs and production rates. Permitting for both water supply and effluents can be costly and time-consuming. Some states, such as Maryland, are trying to streamline the approval process to assist aquaculture development. Recirculating systems are particularly competitive in areas with water shortages or expensive permit requirements. These systems, however, also require diligence in operating and maintaining water-quality conditions to optimize growing and avoid catastrophic crop losses. Open-water pens and floating cages, while taking advantage of natural space and water resources, must contend with other demands such as navigation, recreation, and aesthetics. Finally, in some areas it makes sense to use brackish water supplies. While it may inhibit other water uses, brackish conditions may be ideal for certain types of aquaculture, such as shrimp culture.

The location and characteristics of markets also influence the aquaculture habitat. If an aquaculture producer in the tropics targets nearby resort areas, its territory is regional. However, if the same enterprise targets northern markets, the territory becomes interregional, with products frequently transported thousands of miles. Shrimp that are raised in Ecuador, for example, may be shipped to the mid-Atlantic U.S. and then air-freighted to Colorado. Such transcontinental commerce, which is common in our industrialized society, is dependent on cheap energy and transportation. If the target customers prefer fresh seafood grown locally, the producers may choose to locate closer to the market.

An important drawback with tropical and subtropical aquacultural production is the distance to the markets in the north, where customers can and do pay premium prices for shrimp and other warm-water products. The savings resulting from the warm climate and cheap labor in these regions are offset by transportation and preservation costs. The fish tilapia, for example, is beginning to be grown closer to these markets, such as metropolitan Boston, where fresh products command even higher prices with upscale restaurants and grocery stores. Tilapia, a fish originally from Africa and the Middle East, is popular as a live whole fish in Asian communities within the large urban centers of the north. It is also presently in vogue among health-conscious shoppers at large health food stores such as the chain, Whole Foods Market®.

Living resources, as shown in Figure 3.2, must be available to aquaculture producers. Certain species may be in short supply because they are difficult to breed and raise, or they are susceptible to disease. In other instances, some species, such as sturgeon, may not be available because they require several years to reach maturity. By comparison, tilapia require only several months to reach maturity.

Biological requirements of aquatic animals and plants affect the aquaculture habitat in a number of ways. Important characteristics affecting the health of aquatic animals and plants include water quality (such as temperature and dissolved substances), physical characteristics (such as tank or open water, lighting, and substrate), and ecological characteristics (such as size, density, type, and distribution of animals and plants).

If, for example, a species is hardy and tolerant of changing conditions, open-water culture in pens or pond culture may be feasible; striped bass and tilapia are examples. However, the climate will dictate the growing and harvesting season unless closed environments, such as greenhouses or plastic domes, are created. Kelp is being raised and harvested off the coast of California, where it grows naturally, as a potential source of food, energy, fiber, and chemicals. Shrimp, which can be grown in open and closed systems, are dependent on a warm-water environment that can vary from fresh water to salt water, depending on the species. Finally, duckweed, a floating aquatic plant, thrives in the partially treated effluent from aquaculture and water reclamation. It is remarkably similar to soybeans in nutritional content and provides a low-cost food or food supplement for farm-raised fish and livestock. According to *Duckweed Aquaculture,* a 1993 publication of the World Bank, duckweed will become a significant source of protein, vitamin A, and pigment once commercially viable drying and extraction technologies are developed.

The rate and quality of the production environment affects the health of aquatic animals and plants, and consequently the quality and marketability of products and services. According to John Reid of Bioshelters, Inc., his company has produced fish for more than ten years without the use of antibiotics. Reid says that by minimizing the stress on the fish, disease control measures are not necessary. He adds that the industrial philosophy of maximizing production, regardless of the stress imposed, is common within the poultry industry, and is making its way into the aquaculture industry as well. Disease outbreaks and fish kills often occur because aquaculture facilities are pushing the biological limits, rather than seeking healthy, balanced, optimal levels of production. By not using antibiotics, firms such as Bioshelters, Inc. are positioning themselves to capture the expanding, health-conscious market.

Labor costs and availability of well-trained technical and business personnel are vital to the success of an aquaculture business. Highly skilled technical personnel are required since considerable biological knowledge is often required, as well as training in chemistry, engineering, and computers. This is particularly true in the large-scale, recirculating aquaculture systems that, in many ways, resemble natural ecosystems. Problems, such as disease and water impurities, disperse quickly through such systems if they are improperly maintained and monitored. The computer-based monitoring systems that have been developed to address these problems resemble the self-regulating systems of organisms or ecosystems.

Labor is particularly important in the harvesting and processing stages of aquaculture. The filleting of tilapia, the shucking of oysters, and the packaging of aquaculture products, for example, are all labor-intensive activities. Aquaculture that is focused on ecological restoration can also provide employment to individuals and groups who have suffered economically from environmental degradation or overharvesting. Underemployed watermen, for example, can be trained and hired to help replenish depleted fish and shellfish resources. High school and college students are also a valuable resource for aquaculture enterprises who, in turn, can provide budding entrepreneurs with practical internship experience.

Business expertise, finally, is required to ensure the successful development of an aquaculture business. Problems experienced by aquaculture enterprises—particularly the smaller ones—stem from a lack of business knowledge, marketing experience, and sound financial planning, and the perception on the part of the investment community that aquaculture is "risky" business. The perception of risk is often due, in part, to the fact that many lenders are unfamiliar with aquaculture because the industry has yet to develop a lengthy track record of financial success. Consultants and government programs such as the U.S. Small Business Administration can often provide technical assistance in the areas related to business development (see consultants in Figure 3.2). In addition, schools, such as St. Mary's College in St. Mary's City, Maryland, are developing formal training programs.

Two Examples of Aquaculture Shaped By Social DNA: Family-Owned and Industrial

How does social DNA—an organization's core purpose and values—affect the organization and development of aquaculture? To answer this, it is useful to compare different types of aquaculture, particularly in the context of sustainable enterprise. In this case, we will examine traditional, family-owned microenterprises, which link the local economy, culture, and environment, and large-scale industrial systems, which tend to maximize production and profits for shareholders at the expense of other stakeholders and the environment.

Traditional aquaculture systems in Asia blend the production of fish and shellfish with other farming practices, such as rice and hog production. Chapter 5 highlights these "agro-ecosytems" as an important learning model for business ecosystem development. T. V. R. Pillay, a recognized aquaculture expert from India, describes this form of aquaculture and the subsequent impacts of "modernization" in the June 1996 issue of *World Aquaculture*:

> Traditional aquaculture, especially in the developing countries of the Southern Hemisphere, evolved as small-scale peasant enterprises to suit rural economies. This type of aquaculture has the following characteristics:
>
> - family ownership;
> - polyculture;
> - integration with crop and animal farming activities;
> - waste recycling and beneficial use of farm wastes;
> - diversification of food production and spreading of farmer's risks;
> - provision of off-season work for farmers and wage-earners; and
> - a general means of improving nutrition and incomes.
>
> Traditional aquaculture was therefore considered eco-friendly and was promoted by development agencies and governments. What has happened to change this perception? In a bid to increase the production and profitability of aquaculture, and more importantly to "modernize" operations (so as to shed the appellation "traditional"), production systems have been intensified without considering the consequences. Intensified "modern" culture systems appeared economically remunerative, and as export opportunities opened up, individual entrepreneurs, large companies and corporations entered the field. Aquaculture development agencies were pleased with the flow of capital and industrial management expertise into the sector, but in many areas the small "traditional" farmers felt threatened. This inevitably led to social conflicts which often assumed crisis proportions, depending on the intensity of local politics and the degree of environmental activism.
>
> Cooperation between the Southern Hemisphere and the Northern Hemisphere has played an important role in the so-called modernizing and intensifying of culture techniques in the developing regions of the world. Even though aquaculture production in the North is still less than 12 percent of the total, northern institutions and technicians have contributed substantially to the modification of farming technologies

of the South, especially in the production of inputs such as feed, seed, therapeutics, and farm equipment.

Unfortunately, these new technologies brought with them many of the risks involved. Even though the same problems are faced in the countries of the Northern Hemisphere, these countries appear better able to deal with them, possibly because of their greater scientific support and organizational advantages. Even if further expansion of aquaculture were severely restricted or even banned, I believe the North would suffer less than the South, especially in Asia where more than 85 percent of cultivated fish and shellfish are produced.

Conflicts over the multiple use of natural resources are generally of a different nature and magnitude in the North and South. The conflicts in the North are typically recreational users, real estate builders, wetlands conservationists, and seafront dwellers who treasure the scenic view from their homes. In the South, particularly in Asia, the problems are more of an environmental and social nature, including:

- salination of soil in agricultural lands, due to seepage from coastal aquafarms;
- diversions of agricultural land and mangrove forests for aquaculture use;
- salination of drinking water sources in neighboring villages;
- competition in the use of underground water;
- displacement of small farmers;
- obstruction of free access to the waterfront and subsistence fishing; and
- deterioration of fragile ecosystems.

This passage describes two distinctly different forms of aquaculture, a traditional family-owned microenterprise and a large-scale industrial system, and illustrates how the needs of the North and South differ.

Traditional aquaculture links the local economy, culture, and environment. It is a source of food and livelihood for the local populations. The systems of production are tied directly to the natural resource systems, creating closed-loop economies where by-products from one activity become inputs to another. Local people also have a stake in the success of local microenterprise in terms of food and employment.

In contrast, the prevalent industrial systems are driven by short-term earnings gained through production maximization, with little or no concern for the environment and the local economy. Pillay describes how entrepreneurs who invest in aquaculture in the Southern Hemisphere are "less concerned about sustainability than about immediate return on investment." They apply a linear, factory model that considers how inputs and outputs can maximize earning streams that satisfy investors: extract, deliver, produce, market, and sell. The carrying capacity of the local environment is of little concern, because it is assumed that once the profits are made, the enterprise can move on to extract resources somewhere else.

As introduced earlier in this chapter, aquaculture holds much promise as a means for feeding a growing global human population. It can also be an important source of pharmaceuticals, biotechnologies and biomaterials, pollution control and ecological restoration, energy, fertilizers, bait, and ornamental plants and animals. A key to sustainable development of aquaculture is establishing effective partnerships between the Northern and Southern Hemispheres and being sensitive to the needs and cultures of communities in both regions. Fostering a genuine spirit of community, where the values of all stakeholders count, can help ensure a sustainable development future for aquaculture.

Many governments of the South, according to Pillay, are also creating business environments that favor export trade and earning foreign exchange over national goals of self-sufficient food production and meeting local nutritional needs. But this sort of development is not sustainable because it undermines the local economy and communities. Pillay recommends the development of "appropriate methods" of production, that reflect local conditions and the "carrying capacity" of natural resources:

> The blind pursuit of intensive production should be replaced by "appropriate methods" of production, which may include the whole spectrum, from the so called "extensive to super-intensive," provided they are suited to local environmental and socio-economic conditions. It may be possible to attain sustainability by simple adjustments to the decisive elements of any of these systems according to local requirements.

"Carrying capacity" of the water body may be a determining factor in open-water farming, as in coastal areas, but in enclosed farms appropriate technologies to achieve higher levels of sustainability may require suitable combinations of the following in order to obtain optimum economic and social benefits without damaging the environment:

- adjustment of stocking rates;
- adjustment of feed quality and feeding rates;
- effective water management;
- avoidance of losses due to diseases;
- control of effluent quality; and
- adjustment of harvesting time and size at harvest.

Pillay's account illustrates how Western development policies created conditions for "modern, intensified techniques" to flourish at the expense of more "traditional" approaches. It also illustrates what happens when stakeholder relations are not considered.

An aquaculture enterprise creates value by combining life-sustaining flows, such as money, energy, food, and water, to cultivate commercially viable, value-added products and services, which, in turn, are marketed, sold, and distributed to customers. Important factors affecting its metabolism, niche, and habitat include: climate and environmental considerations; access to transportation, space, energy, food, water, and nutrients; location and characteristics of markets; biological requirements of aquatic animals and plants; labor costs; and availability of well-trained technical and business personnel. **Value creation, efficient use of life-sustaining flows, and building healthy stakeholder relations, as the examples show, are vital to sustainable aquaculture development.**

As the examples in this chapter illustrate, business ecology provides a lens for seeing vital relations that sustain your company within its environment. Once you begin examining the vital flows and relations among the organism, its stakeholders, and its environment, you will see new opportunities for efficiency and the attainment of your organizational goals. Your company's life-sustaining flows, together with its stakeholder community, determine your company's "environment," including: its metabolism; its internal economy; its niche, how it makes a living; and its habitat, where it lives.

Thinking of the living organism as a simple, elegant, organizational model that can be applied to organizations of every scale and type, opens a new field of vision for business and organizational management. Through business ecology, interrelated metabolic processes—values-based organization and development, value creation, and cyclical flows—are viewed as interdependent, fluid, self-organizing systems. These systems dynamically interact within your organization and collectively exchange flows with the external environment. Social DNA, as the family-owned and industrial aquaculture examples demonstrate, are the core values that determine how your company organizes and develops.

The ultimate goal of business ecology and sustainable enterprise is to efficiently use life-sustaining flows to produce high-quality, value-added products and services. The next chapter looks more closely at each of these life-sustaining flow systems and describes how this systemic vision can improve your organization's or business's viability and help you gain competitive advantage in the ecological economy.

4

Looking Beyond Cash Flow

If your only tool is a hammer, you see every problem as a nail.
Abraham Maslow

An ecological perspective of your organization can transform the way you think about accounting, management, and economic development. In this chapter, we will take a holistic look at the flows that sustain your business or organization. The flows that define your company's metabolism, niche, and habitat include: people and other organisms, food, information and ideas, money, products and services, air, energy, water, and materials. Why is each flow important to your business or organization? How does the relative importance of various flows—such as food in the restaurant business or information and ideas in the publishing industry—define metabolism, niche, and habitat in different sectors and industries, including how vital flows may be exchanged? For instance, is your company a net producer or user of energy? Are there by-product materials produced by your company that could be recovered internally or sold to another company? Are you efficiently closing the loop on vital flows? Tangible benefits of "looking beyond cash flow" include increased revenues and profitability, new markets, lower liabilities, pollution reduction and prevention, higher motivation and productivity, and healthier stakeholder relations. In fact, more comprehensive accounting of life-sustaining flows provides a better picture of your company's metabolism and its environment. **What is essential to overall planning is that these vital resources are often limited, or in other cases, underused and undervalued.** To help you understand the vital flows within and outside your company, this chapter provides different views of business metabolism:

- Your Organization's Life Force (people and other organisms)
- The Intelligent Organism (information and ideas)
- Powerful Insights (energy)
- Water: *The* Liquid Asset (water)
- Creating a Healthy Atmosphere (air)
- We Are What We Eat (food)
- Discovering the Lost Economy (materials, products, and services)

The last section, Closing the Loop: How Systems Thinking Can Help Your Bottom Line, describes how companies are improving their profitability, and cash flow, by:

- Seeing and improving their company's metabolism
- "Closing the loop" with other companies
- Developing new products and services from residual flows
- Creating multiple profit cycles

YOUR ORGANIZATION'S LIFE FORCE

Viewing your company as an organism reveals what thriving companies and true leaders have always known—an organization's most valuable asset is its life force. This includes people and other "organisms" that support a company's metabolism. These companies develop long-lasting relationships with their stakeholders, because the stakeholders' satisfaction, trust, commitment, and financial and technical support are crucial to their organization's success. Key stakeholders usually include: customers, employees, shareholders, creditors, suppliers, manufacturers, distributors, utilities, retailers, citizen and community groups, various levels of government, contractors, accountants, and lawyers. Nonhuman life-forms can comprise a significant portion of an organization's life force. Examples include cattle in the beef and dairy industries, sheep in the wool industry, trees in the forest products industry, yeast in the brewing industry, silkworms in the silk industry, and medicinal plants in the pharmaceutical industry. Even petroleum, natural gas, and other "fossil" resources were made available to us through the work of organisms that lived long ago. Collectively, all these "organisms" comprise a company's food web and ecosystem.

One of these—employees—brings the ingenuity, energy, and creativity that ultimately makes a company successful and viable. Healthy and motivated employees who have a sense of purpose, and the right recognition and resources, can accomplish amazing things, including higher profits and sustained success. This form of success exudes life throughout a company and its environment, affecting customers, suppliers, and other stakeholders. Malden Mills, the makers of Polartec® in Lawrence, Massachusetts exemplifies a profitable, community-based company that sees its people as its most valuable asset.

Aaron Feuerstein, owner, president, and CEO of Malden Mills, gained national and local praise for his response to a December 11, 1995 catastrophic fire that devastated 750,000 square feet of factory space and left nearly 1,400 people (about two-thirds of the workforce) with no job to go to the next morning. In a period when the U.S. textile industry was chasing cheaper wages in the southern U.S. and offshore, he decided to rebuild the factories and provide pay and health benefits to 1,400 nonworking employees during the first few months of initial recovery. Later, he would also assist workers that he was unable to hire back to find new jobs until he could bring them back. His leadership and generosity generated a massive wave of aid from public and private sources, including local businesses, neighboring towns in northern Massachusetts and southern New Hampshire, the Commonwealth of Massachusetts, the U.S. government, and countless benefactors and customers from across the U.S.

While he was praised by many citizens, leaders, and groups across the U.S., there were those who questioned his business acumen. Many business "experts" believed that cashing in on the insurance and retiring or rebuilding in a region of cheap labor were the obvious, "smart" business choices. Here is Feuerstein's answer to these bottom-line, shareholder-driven "experts," as printed in the *Journal of Innovative Management* (Fall 1996):

> The fire that we had was really devastating. Three major mills, 250,000 square feet each, were totally destroyed to the ground, nothing left. As CEO, I had the responsibility at that time to make a decision, and I made it: To rebuild as fast as humanly possible. And I am asked by professionals and by some of the modern CEOs, why did I do it? According to them, I really wasn't doing what was best for the shareholder, and the shareholder is king today. Perhaps I could have made a much better deal with the insurance company, and cashed in all the stock and all the hard labor and sweat that went into three generations of work.
>
> There are two answers to that kind of criticism: One, it's wrong. What we have to do is the right thing; and two, from an economic point of view, if you look at the long-run instead of the short-term, I'm

sure I did the right thing for the profitability of my family, the shareholder. I equally had a responsibility for the workers who really participate with us to make our products the very best in the industry.

Again, they ask, "In what way are you helping the shareholder by
taking care of the worker? You surely can move to other locations
where you can have cheaper labor and you really are not doing what
a CEO is supposed to do?" Well, if you think of the laborer as a pair
of hands, as a fixed expense that's cuttable, then yes, the criticism is
right. However, the philosophy we have at Malden Mills is that the
laborer is an asset, and he participates with us in making our products,
and his loyalty and trust is extremely, extremely important.

Today, Malden Mills is a highly successful company with a new, state-
of-the-art plant, built over the footprint of the three devastated buildings.
It is one of the most technically and environmentally advanced textile mills
in the world. Their global leadership in the textile market is also highlighted in the *Journal of Innovative Management* (Fall 1996):

> While most of their competition competes on price, Malden Mills relies
> on its skilled work force, offers higher quality, has more diverse and
> innovative products, and listens carefully to the needs of its customers.
> They spend much of their revenue on Research and Development, and
> pay their workers an average of $12.50 an hour, compared to an in-
> dustry average of just $9.44. They have moved from $5 million in sales
> of Polartec in 1982 to $225 million in 1996. Their total sales are over
> $400 million a year, with customers in more than 50 countries. Their
> number of employees peaked last year at 3,200, and they expect to hit
> sales of $1 billion by the year 2000.

Clearly, building long-lasting, responsible stakeholder relationships
can also be good business. Chapter 6 profiles additional organizations that
are reaching higher levels of success, including higher profits, by developing
respectful, enduring relationships with their stakeholder communities.

THE INTELLIGENT ORGANISM

The intelligent, life-sustaining organism will have a definite advantage in
what is becoming a new era of business ecosystems, global consciousness,
purposeful work, and civic responsibility. Information about local market
niches, changes in the business environment, and feedback loops from cus-

tomers and other stakeholders will provide a competitive advantage to twenty-first century organizations. The Internet and other communication media are redefining organizational intelligence and how information is accessed and shared. No doubt, bureaucracies built on restricting ideas and information flow will find it difficult to survive in the open twenty-first century.

For the intelligent organism, surviving means being creative and open to finding "food"—essential, sustaining flows of information and ideas. For example, a magazine publishing company forages for information and ideas to create reading and visual experiences that attract subscriptions and a circulation that attracts advertisements. Fresh ideas and information are "metabolized" through the orchestration of writers, editors, graphic artists, and effective communication with advertisers and the community the publication serves.

Fast Company, a business magazine with an estimated international circulation of 175,000 in 1997, is an example of a "knowledge-propagating business" in the publishing industry. It profiles new thinking, trends, and change agents in the business community. It also caters to the savvy image the reader may have of him- or herself as a creative, maverick business thinker. Whatever the draw for its readers, *Fast Company*, like many magazines, does in fact provoke new ideas and create a community. It is interactive, using the Internet to help its community of readers connect with each other and featured experts. It is synergistic, cross-pollinating ideas and strategies that work among its readership community, and stimulating sales of the products and services it advertises.

Here are highlights drawn from a *Fast Company* article (October/November 1996), written by founding editor Alan M. Webber, that point to the emergence of more organic, intelligent work and "feeding" environments. "Think You're Smarter Than Your Computer?" is an interview with James Bailey, author of *After Thought: The Computer Challenge to Human Intelligence*. Webber describes Bailey's book as a "profound analysis of the history of thought." During the interview, Bailey describes how computers and networks are acquiring organic qualities and how this is reshaping our work environment.

Bailey explains that how we think and process ideas and information is changing, and how this will shape the future of business. He believes that **we need to change the way we think, and learn to use computers and other information technologies to see biological *patterns*.** Bailey describes three revolutions that relate to our evolution in thinking:

1. Geometry and "place"
2. Technology and "pace"
3. Biology and "patterns"

In the first revolution, geometry provided a means for helping people relate to a sense of place: "People in the ancient world wanted to know where they were in relation to the universe. The math they developed to identify place was geometry," says Bailey. This included tools for delineating cities, navigating the oceans, tracking celestial bodies, and exploring and mapping new lands.

Technology grounded in physics, according to Bailey, became the driving force of the second revolution, which is the dominant mode of thinking today. The world is characterized in terms of numbers, equations, and objects. This is the mechanistic, industrial-age worldview. He suggests that clocks, for example, made people more aware of "pace." The printing press and subsequent information technologies made text and numbers available to more people and quickened the dissemination of ideas, information, and knowledge.

Biology and self-organizing systems are the basis of the third revolution, with a focus on relationships and patterns. Bailey's new ideas on "how we think" is rooted in his work with Danny Hillis who founded Thinking Machines Corporation more than ten years ago. Thinking Machines' Connection Machine is one of the world's first large-scale, parallel-linked computer networks. Such a network enables computers to work together like cells in the human brain: they can solve complex problems while increasing their problem-solving capacity with each new solution. In other words, the network "learns" and gets "smarter." Webber's article describes Bailey's discoveries as Thinking Machines' marketing director, after he left a comfortable job at Digital Equipment Corporation:

> He immediately confronted two mind-bending realities. The first involved Hillis's machine, which fundamentally changed the logic of computing. It not only accelerated computation, it also processed data in a new way, looking for patterns and learning. In short, it made itself smarter. The second involved experiments with "cellular automata"— bits of information that operate according to a few simple rules. Inside one of Hillis's computers, the cellular automata began to organize themselves and patterns began to emerge. In short, they acted as if they were alive.

Clearly, Bailey's observations and comments are consistent with business ecology thinking, and highlight the significance of information and ideas and the pattern of relationships within organizations.

In view of largely fixed natural resources, and in the face of rising world population and expectations, it is becoming increasingly obvious that humankind must become more ecologically efficient. This means using existing resources more wisely and efficiently. Happily, most approaches that improve ecological efficiency also improve the bottom line. **Many such "efficiency" technologies and techniques, which allow us to reach new levels of productivity, can be optimally managed with information technologies.** The first companies to realize this and act upon the realization will also gain a competitive advantage.

In Painesville, Ohio, and in Pittsburgh, Pennsylvania, for instance, intelligent technology has enabled more efficient combustion within large electric utility boilers. Pegasus Technologies, Inc. invented and is now deploying a new neural-network technology that monitors combustion parameters in boilers for the purpose of making continuous "trimming adjustments" to the combustion. This system detects small suboptimal changes in the combustion process and then corrects them automatically. The system improves the control of the boilers beyond the capabilities of human-only controls. This technology is poised to provide significant environmental benefits to the U.S. By optimizing combustion parameters, less fuel is required to produce the same amount of electricity. Less pollution is generated, including sulfur dioxide and nitrogen oxides, the primary causes of acid rain. In addition, much less of the greenhouse gas carbon dioxide is emitted to the atmosphere, which lessens man's impact on the global climate. Many utilities are now being retrofitted with the Pegasus system because they can achieve large reductions in air pollutants through cost-effective efficiency improvements. Achieving those reductions otherwise would require costly pollution control equipment and downtime.

By infusing creativity, ingenuity, and "market awareness" throughout your organization, information and ideas simultaneously enhance your company's metabolic efficiency, its viability, and its bottom line. Information and ideas, and the ability to see biological patterns, are critical in a knowledge-based, ecological economy. Business ecology, by drawing on natural systems design, can help you assess and develop this flow, and help your organization develop into a self-organizing, life-sustaining, intelligent organism.

POWERFUL INSIGHTS

Access to energy is essential to all living systems. Energy is a driving force of our economy and way of life. However, the energy sources we choose to use can cause a host of environmental, economic, and health problems, sometimes undercutting the very quality of life that inexpensive energy has made possible. Ample opportunities exist to save energy, recycle residual energy flows, use renewable energy, and enhance energy security while allowing people to enjoy the same or an improved quality of life. Why are these energy opportunities important to your organization? First, you can save money today by becoming more energy efficient. Companies such as 3M, Monsanto, AT&T, and Dupont are saving millions of dollars through energy efficiency. Second, the finiteness of fossil fuels and environmental issues (e.g., such as climate change) are already driving corporate behavior. Oil companies such as British Petroleum (BP) and Royal Dutch/Shell are responding to these challenges by positioning for strategic advantage (BP's 1997 announcement to invest $1 billion in the next 10 years in solar energy). Arie de Geus, who is renowned for his strategic planning with Royal Dutch/Shell, writes in his book, *The Living Company*:

> . . . Shell executives cannot avoid discussing the question: Is there life after oil? What other businesses might Shell reasonably enter? How might we prepare for switching to them as our primary business? And what effect would that switch have on our company as a whole?

Considering our dependence on petroleum, how will its limits or shifts to other sources affect your company? Here is a perspective from Ray Anderson, president, chairman, and CEO of Interface, the world's largest carpet tile manufacturer. He described his company's strategic intent and the limits of ecological efficiency within our fossil-fuel economy in *Business and the Environment* (October 1997):

> We want to get end-of-life products back into the supply chain, but it's incredibly difficult to do. We're also trying to figure out how to use natural materials to make compostable products. It does no good to close the loop if you drive the process with fossil fuels.

Anderson considers our fossil-fuel dependence to be a dangerous addiction. He has introduced over 400 projects to transform Interface into a sustainable company, including a push towards renewable energy and energy-efficient transportation. To help you gain an understanding of the potential impact of energy and its use on your business, we will explore four areas of energy production and use: electricity generation, transportation, buildings, and industry. In each of these areas, new technologies combined with the mindware of business ecology can save energy, improve profitability, and enhance our quality of life. The following information is drawn from "Pollution Prevention Strategy for the Energy Sector" (1991), a U.S. EPA report prepared by David Bassett, and updated with 1996 energy data from *Annual Energy Outlook, 1998.*

Electricity

Electricity is a key ingredient in any modern economy. In the U.S., electricity is the most rapidly growing portion of the energy sector; in the postwar era, its usage has grown three times as fast as the economy as a whole. By 1996, 36 percent of the total primary energy consumed in the U.S.—over 34 quadrillion British Thermal Units (BTUs)—was used to generate electricity. The steady growth of electricity use reflects our expanding wealth in electric-powered equipment and infrastructure—tools, appliances, computers, telecommunications networks—many of the items we consider the hallmarks of an advanced society.

As the production and use of electricity have expanded, so have the environmental side-effects. Electricity generation is responsible for a significant portion of national emissions of several pollutants, including about 66 percent of sulfur dioxide, 35 percent of nitrogen oxides, and 35 percent of carbon dioxide. The environmental impacts of electricity use go well beyond emissions from combustion; they include the ecological impacts of surface mining and power line corridors, the long-term risks associated with radioactive nuclear waste, the loss of unique habitats caused by hydroelectric dams, and the vast quantities of water used to cool power plants.

The potential for improving ecological efficiency in the electricity subsector is enormous. In the near-term, district heating and cooling systems, cogeneration, profitable deployment of energy-efficient devices such as those advanced by the U.S. EPA's Green Lights and Green Star Programs, demand-side management by public utilities, improved operation and maintenance of generating systems (see Pegasus Technologies, Inc. in the section entitled "The Intelligent Organism" above), daylighting of buildings, light-colored buildings, and deciduous tree planting are examples of viable options that save money while reducing electricity consumption and its accompanying ecological effects. Certainly, as the Kalundborg, Denmark, business ecosystem demonstrates, closing the loop is an effective near-term solution for reducing environmental impacts while saving money and improving resource use efficiency. In the longer-term, investments in renewable energy technologies, such as wind, solar-thermal collectors, photovoltaics, geothermal sources, and biomass, offer ecologically superior forms of electricity generation that often can be placed closer to the consumer, thus reducing infrastructure costs and vulnerability to centralized system failure.

Transportation

The U.S., with its dispersed population, large economy, and unique geography, requires a diverse and complex transportation system to meet its needs, yet the environmental consequences of our current solutions to the transportation challenge have resulted in considerable damage to human health and ecology. Future demands for transportation are projected to increase steadily, intensifying environmental problems, including air toxins, criteria air pollutants (e.g., sulfur, nitrogen, and carbon oxides), and global climate change.

Transportation uses about a quarter of all energy consumed in the U.S.—over 24.8 quadrillion BTUs. The environmental implications of reliance on petroleum-based transportation technologies—especially the internal combustion engine—are considerable. Transportation accounts for about 70 percent of emissions of carbon monoxide, 43 percent of nitrogen oxides, 31 percent of volatile organic chemicals, and 32 percent of carbon dioxide. Much of this pollution results from energy inefficiencies in moving vehicles. For every gallon of gasoline consumed about 60 percent of fuel is lost as waste heat; 15 percent to 20 percent is lost to overcoming internal engine friction and to power accessories; 5 percent is consumed by driveline friction; 10 percent is lost to aerodynamic drag; and only about 10 percent provides the transportation service. Much of this latter amount is ultimately dissipated as brake heat.

A number of technologies and options are available for improving transportation efficiency and reducing its ecological footprint. These include decreasing vehicle weight, improving combustion and transmission efficiencies, using energy-conserving braking systems, and developing alternative power systems for the internal combustion engine, such as a fuel cell/electric motor and solar/electric motor systems. Alternative, cleaner-burning fuels for gasoline and diesel include methanol, ethanol, compressed natural gas, and hydrogen. However, **the overall viability of these fuels de-**

pends on their life-cycle costs, including the environmental consequences of their production and use and whether they can be derived economically from renewable sources.

Transportation system management includes both demand and supply-side improvements in efficiency. Reducing vehicle-miles-traveled through better planning and telecommuting are important demand-side options that provide obvious cost savings and social and environmental benefits. Incentives or disincentives to get people out of their cars is another. These include pricing strategies for highways, mass transit, parking, alternative working arrangements, public education, and land-use planning. Use of mass transit, bicycles and bike paths, and high occupancy lanes are examples of supply-side options. In general, most alternatives to auto-based transportation are more environmentally benign.

Economic barriers to using available alternatives are considerable, reflecting a key value of the industrial economy—subsidized, cheap, fossil-fuel-based transportation. Prices do not reflect the true costs incurred in vehicle transportation and are therefore artificially low. For example, consumers generally do not consider the consequences of fuel efficiency over the life of an automobile when purchasing new cars, so the demand for high fuel-efficiency remains low. Consumers also do not bear the full costs of using highways or city parking. Parking alone costs between $1,000 and $15,000 per space to construct, plus maintenance. More than half of highway expenditures come from general tax revenues as opposed to user taxes and special taxes. If the costs borne by general tax revenues were incorporated into the costs of the fuel or vehicles, consumers would likely consider more benign transportation modes and alternatives.

Buildings

Energy consumption in both residential and commercial buildings represents nearly 20 percent of total energy consumption in the U.S. A total of 18.6 quadrillion BTUs are consumed annually, with an estimated 11.55 quadrillion BTUs consumed by direct combustion of fuels for space and water heating, and 7.05 quadrillion BTUs of electric energy consumed by air conditioning, refrigeration, freezing, appliances, and lighting.

The building subsector is the largest, single end-user of electricity in the U.S., accounting for two-thirds of total electricity consumption. Lighting alone consumes approximately 25 percent of this electricity—20 percent directly and 5 percent in cooling equipment to compensate for unwanted heat from lights.

Considerable quantities of the energy consumed in buildings are lost each year through combustion inefficiencies and inefficiencies in building construction, maintenance, and operation. In many cases, the economic costs of energy use (and loss) are not borne by builders, and consumers often do not have the ability or the incentive to influence the efficiency of the buildings in which they live and work. **Only recently have industrial consumers become aware of the costs associated with energy inefficiency.**

The environmental consequences of energy use in buildings are sizable: 23 percent of all domestic emissions of carbon dioxide, 17 percent of nitrogen oxides, 15 percent of volatile organic compounds, and 14 percent of sulfur dioxides. As in other subsectors, substantial opportunities exist to reduce emissions while realizing substantial savings to the individual consumer and society.

Initiatives to reduce energy demand by the building subsector overlap with many of the demand-side management programs discussed in the previous section on electricity. Thus, commercial and residential consumers in existing buildings can take a variety of energy-saving actions, such as improved lighting efficiency, that will reduce energy demand and environmental emissions.

More permanent and reliable changes in the demand for energy from the building subsector need to be incorporated in new buildings. That can only happen by reaching architects, building planners, city officials, businesses, and prospective homeowners. Near-term initiatives can seek to change behavior and encourage the construction of more environmentally sound buildings. Business ecology, as a framework for transforming organizations into sustainable enterprise, can help accelerate the shift to more ecologically efficient buildings by highlighting bottom-line and stakeholder benefits, such as lower operation and maintenance costs, a healthier work environment, and higher productivity.

Industry

American industry consumes about 23.6 quadrillion BTUs of primary energy and contributes over 49 percent of volatile organic compound emissions, 15 percent of sulfur dioxide emissions, and over a quarter of carbon dioxide emissions. Additionally, industry consumes over 3.4 quadrillion BTUs of electric energy derived from some 11.2 quadrillion BTUs of primary energy. **Most of this energy is consumed in manufacturing products, but a significant portion, estimated at over 3 quadrillion BTUs, is locked** up as "embedded energy" in products themselves. Examples of this are abundantly found in products such as glass, steel, and aluminum, which represent the energy invested in refining raw materials. To make an aluminum can from bauxite, for example, 100 energy units are required. To make the same can from recycled aluminum metal, only 5 energy units are needed. Clearly, closing the loop on materials can have positive efficiency improvements for a business or organization in other flows such as energy, water, and air.

A number of opportunities exist for savings in industrial energy use with associated benefits to the environment. Energy conservation measures can significantly reduce carbon dioxide emissions by cutting down on the amount of energy used in industrial processes. Small amounts of energy can be conserved at low-to-negligible costs by changes in basic "housekeeping" practices. Improved "housekeeping" methods, such as turning equipment off when not in use, are economically desirable and appeal to common sense. Other opportunities that exist in this area include better lighting systems, more efficient variable speed motors and controls, and more efficient recovery of waste heat.

Major opportunities lie in the improvement of alternating current motors and associated drivetrains. Motive power is the largest use of electricity in the industry subsector. At present, 40 to 80 percent of the total motor energy used is lost because of inefficient speed control. Modern electronic controls and efficient drives can significantly increase electricity savings and overall energy savings. Using currently available technology, about 30 percent of industrial energy can be saved, with an associated reduction of 140 million tons per year of carbon dioxide emissions.

Here is an example of a company whose business is saving energy, cutting costs, and reducing pollution. Trigen-Trenton Energy Company, L.P., based in Trenton, New Jersey, "makes its living" by providing energy efficient, closed-loop solutions to communities, businesses, government agencies and residential clients. Trigen stands for "trigenerated" heating, cooling, and electricity. Here is a description from the company:

> Trigen-Trenton leverages two concepts, *district heating and cooling* and *cogeneration,* to minimize environmental impact, reduce user costs, simplify building operations and ensure dependability. Cogenerated district heating and cooling (CDHC) allows many energy consumers to be served by one supplier using more efficient equipment and techniques, and it doubles the productivity of each unit of fuel burned.

Trigen-Trenton is supporting Trenton's continuing revitalization (see Chapter 4) with a "modern, environmentally green 'backbone' of energy." According to Vice President Donald Lebowitz:

The Trigen story really gets to the heart of society's current concerns, namely, energy conservation, competitiveness, and the environment. All of these considerations are inherent benefits of district heating and cooling and of the Trigen Trenton project in particular.

Energy conservation is achieved by reducing fuel usage. The Trigen project recovers engine exhaust by transforming it into high-temperature hot water and steam. It further makes use of engine jacket water heat in a separate, low-temperature loop that heats apartment houses. The system reduces the amount of oil and gas consumed by a total of nearly 50 percent; part of the savings comes from recovering heat from electric generation that would otherwise have been wasted.

There is also increased boiler efficiency obtained by transferring steam production from inefficient boiler plant systems, capable of only 50 to 65 percent seasonal efficiency, to Trigen's central boilers with annual efficiencies above 85 percent. If district heating and cooling were universally used in our urban centers throughout the nation, significant inroads would be made on the U.S. trade balance, because a major portion of U.S. imports is fuel oil, used to wastefully heat urban buildings. To sell energy, the system has to be competitive, and this saves everyone money, lowering the price of their services.

Another benefit of Trigen's district heating and cooling project and district heat in general is to the local and global environment. Given the reality that a cogenerated district heating and cooling system uses only one half as much fuel as separately generated heat and power, there is obviously at least a 50 percent reduction in carbon dioxide and typically a reduction of carbon monoxide, sulfur dioxide and particulates. Because Trigen's gas-fueled cogeneration replaces some #6 oil-fired boilers, the production of acid rain producing sulfur dioxide is nearly eliminated from the target users.

The local environment also is dramatically improved due to the district heating and cooling plant replacing a large number of small, often unattended, unmonitored boilers with no emission controls and short exhaust stacks. These in-building boilers (17 boiler plants replaced so far in Trenton) were forced to sometimes operate at very low load, which caused excessive smoke emissions. The old replaced boilers lacked variable-speed drive fans or oxygen controls. As a result they exhausted at rooftop level so that much of their pollution reached ground level before it was diluted by mixing with the atmosphere.

By contrast, the Trigen plant's one, 160-foot-high stack emits a plume of much cleaner gas (monitored 24 hours per day for NOx, particulates, and CO). The plume travels above the ground for a long distance, mixing with the atmosphere before reaching the ground. By then the levels of pollutants are well below those designated by U.S. Environmental Protection Agency regulations.

Sustainability implies that energy and other natural resources be used at a replenishable rate, and that these economic flows be cyclical and "in sync" with natural systems. Ultimately, as the true costs of current practice are factored into decision-making, market forces will help new technologies become more available, while changes in our social, economic, and political cultures will make sustainable practices more acceptable and, therefore, possible. Business ecology can help you see and develop sustainable opportunities for creating value—and increasing profits—by using less energy, recovering residual energy from other companies, and tapping renewable energy flows.

WATER: *THE* LIQUID ASSET

Water is essential to life. We use it for drinking, cleaning, cooking, sanitation, cooling, and irrigation. Because it is the universal solvent, it is used in nearly every sort of industrial process or commercial enterprise. Water supply and waste-water services are essential to economic development. But with all these invaluable qualities, water is often taken for granted until it becomes scarce or polluted. Worldwide, disputes over water resources, like those over oil and natural gas, are a great cause of conflict, especially in areas such as the Middle East and Africa. Businesses and economic development groups are realizing that water is, in fact, the ultimate liquid asset.

The Aquarina Country Club in Melbourne Beach, Florida, for instance, considers aquatic ecosystems as integral to their business. In fact, according to General Manager Jim Bates, water and related ecological resources are considered valuable assets for a number of reasons.

First, because the golf course is located in the Saint John's Water Management District, it uses treated, reclaimed water to irrigate its greens and fairways. Bates understands the predicament common to many areas of Florida. The State's growing population and development is depleting the available water supplies from underlying aquifers. While his golf course is restricted from using higher-quality water, he readily accepts that other uses, such as drinking water, should have a higher priority, especially when it serves patrons of the golf course. And since higher-quality water is more expensive, using treated reclaimed water also saves money.

Second, storm-water collection lakes are scattered throughout the golf course, create an inviting habitat for a number of creatures. The littoral zone, the shoreline area where small fish and other organisms tend to congregate, is a favorite feeding area for heron, ducks, and other water fowl. Because these areas are open, the birds and other aquatic animals are less susceptible to predators stalking from overhanging or adjoining vegetation. The Audubon Society has designated Aquarina as a protected waterfowl habitat. Third, observing wildlife and natural resources can be a key draw for many golfers. This is particularly true for people vacationing in areas where they see unusual and beautiful natural scenery.

International Golf Management, Inc. maintains Aquarina and about twenty other courses in Florida. To reduce the adverse impacts on the fragile aquatic ecosystems, they apply pesticides and herbicides only in areas of infestation. Years ago, it was common practice simply to spray the entire golf course.

According to Bates, groups like the National Golf Institute, the National Association of Golf Course Superintendents, and the Professional Golf Association are making significant strides to develop golf courses sustainably. Several courses maintained by the State of Alabama, for example, are designed for and built into flood plains. These courses have raised greens and other features that allow playing to commence shortly after a flooding event. Since other types of development are restricted in these areas, golf courses have found it natural to co-inhabit these areas with valuable riparian wetlands.

Water utilities in the United States have demonstrated economically viable strategies for water conservation. These demand-side management strategies are particularly useful when both water and energy are conserved. Here are some examples from *Industrial/Commercial Drought Guidebook for Water Utilities* (June 1991):

> The city of San Jose studied 15 industrial sites that had implemented conservation measures. Actual water savings per company ranged from two million gallons per year (mgy) to 470 mgy and 25 to 90 per cent of previous use. The 15 firms collectively saved over a billion gallons of water per year worth $2 million per year in water, energy, and sewer costs. Payback periods for the conservation methods were usually less than one year.
>
> The city of Phoenix established a comprehensive Industrial Business and Government program, including some 60 site visits and audits. The potential water savings is about two billion gallons per year.

The Massachusetts Water Resource Authority conducted 34 water audits and 12 case studies. Savings in the range of 20 to 60 percent resulted through measures with a payback of less than three years.

(California's) largest industrial water users include petroleum refineries. Large amounts of water are used for cooling. East Bay Municipal Utility District (EBMUD, Oakland California) plans to supply five million gallons a day of reclaimed water to Chevron's Richmond refinery.

Water is vital to economic development. Limited water resources, in fact, stimulated the development of the Kalundborg, Denmark, business ecosystem (see Chapters 1 and 4). Industries located in the Kalundborg Region are big consumers of water. This region of Denmark, which includes the five municipalities of Bjergsted, Gørlev, Hvidebæk, Kalundborg, and Tornved, has very limited ground-water resources. Enterprises in the Kalundborg ecosystem have cut back on their use of ground water and reuse as much water as possible. Municipal water from Tissø, a regional lake, has replaced 90 percent of the ground-water consumption by the Asnæs Power Station. As the anchoring organism of the complex, the Asnæs Power Station has also reduced total water consumption by 55 percent by reusing its own water and by using both cooling and treated water from the neighboring STATOIL refinery.

At a global level, supplies of potable water are increasingly scarce as population increases. There is a critical linkage between food production and water availability. J. E. Cohen reports that, **annually, about 1300 to 8300 cubic meters of water per person are required to grow food.** According to Cohen, twenty countries in Africa and the Middle East, home to more than 131 million people, had less than 1,000 cubic meters per person available during 1990. Climate change, pollution of ground water and surface water, and shifting development patterns will only exacerbate these shortfalls.

The Monsanto Company, as discussed in Chapter 7, sees water and food as strategic resources for our future. In fact, Monsanto formed both a water and a food production team as part of their seven-team sustainability strategy. Cohen's research highlights the critical linkage between water availability and food production. Water is also essential for sanitation, health, and innumerable social and commercial purposes. Several factors point to the price of water increasing in the future. These include the fact that most accessible fresh water supplies are already developed and there is a reduced availability and quality of existing supplies because of pollution, such as pesticides and nitrates in ground water. Additional factors include: overexploitation, such as the overdraft of the Ogallala Aquifer in the Great Plains of the U.S.; potential impacts from climate change and population increases; growing privatization of water utilities; and decreasing financial and political support for big public work projects. It is clear that water is an indispensable asset for businesses and organizations that will only become more valuable in the future.

CREATING A HEALTHY ATMOSPHERE

Creating a healthy atmosphere in both indoor and outdoor environments is smart business. It is common sense that ample sunlight and fresh, pollutant- and allergen-free air is linked to higher employee productivity and morale. In many cases, pure indoor air quality is essential for successful product manufacturing. In other instances, degradation or modification of the Earth's atmosphere and climate is affecting how organizations, such as international banks and insurance companies, look at long-term risks and risk coverage, and how they invest in energy-related technologies. Clean, clear, fresh air is essential to the tourism industry, especially when customer satisfaction depends on scenic views and outdoor activities. Air is a significant resource supporting fossil-fuel-based industries (such as electric power plants) and transport (such as tractor trailers, air freight, ship, and commercial airlines). It is an input for combustion and a sink for combustion by-products. In the industrial economy, all of these services are undervalued: air is considered a "free good" while social costs such as acid rain and smog are inequitably distributed as "externalities" of the market. Closed-loop systems, as exhibited by the Kalundborg, Denmark, community, are an effective near-term strategy for recovering resources, reducing emissions, and lowering operating and maintenance costs. Here are examples that show why air is a vital flow to any company's metabolism and that creating a healthy atmosphere is good business.

E. L. Foust Co., Inc., for instance, has a simple, but important mission: "Since 1974, the air you breathe is our business." This small manufacturer of high-quality air purifiers owes its origins to the work of Dr. Theron Randolph, who in the 1950s gained recognition as "the father of environmental medicine." Dr. Randolph ran a clinic in Chicago to treat individuals suffering from allergies. One of his patients was Mrs. Foust, wife of E. L. Foust.

When Mr. Foust first met Dr. Randolph, he was a salesman in the steel industry. With the advice of Dr. Randolph, Mr. Foust developed and tested a number of steel fabricated air purifiers from 1964 through 1974 to help his wife's allergic conditions. What began as a labor of love eventually became a commercial enterprise in 1974, when E. L. Foust started his company.

Built using high-quality steel fabrication, the devices are free of plastics and glues that may cause allergies. From its beginning, E. L. Foust has helped people work through their indoor air problems. The company is service-oriented, directing people to information and other businesses even when there is no direct commercial benefit. The result: their customers are their best marketing agents.

Industries such as biotechnology and microchip manufacturing, require ultra-high-quality air and are examples of how total quality environmental management directly impacts the bottom line. Even the slightest pollution or dust can create expensive problems in their ability to "make a living."

At the global level, European-based insurance companies, such as Swiss Re and General Accident and Life Assurance Corporation of Great Britiain, and banks, represented by British Banker's Association, are lobbying governments heavily for measures that curb greenhouse gas emissions. Why? They are convinced that changing climatic conditions are adversely affecting their bottom line. Mark Herstgaard, international journalist, describes the situation in *Perspectives on Business and Global Change* journal (1996):

> And as it happens, two of the wealthiest, most powerful players in the world economy—the global insurance and banking industries—are now coming to believe that their self-interest is incompatible with humanity continuing to pump six billion tons of carbon dioxide and other heat-trapping greenhouse gasses into the atmosphere every year.
>
> The financiers have come to realize that they have at risk literally trillions of dollars worth of property and long-term investments. Most of the beaches on the East Coast of the United States could be gone within twenty-five years, according to a recent estimate by the Intergovernmental Panel on Climate Change (IPCC), a group of 2,500 scientists commissioned by the United Nations to study the problem. With more than $2 trillion worth of insured assets along the U.S. coastlines alone, the threat to the insurance industry is obvious.
>
> Thus, while world governments dawdle, some of the leading banks and insurance companies in Europe and Asia may soon initiate a massive shift of international investment flows—away from fossil fuels and towards solar energy—that will transform not only the global warming issue but the international economy of the next century.

While some industries are still balking at air quality standards, others are strategically positioning themselves with lower- or zero-emitting alternatives that directly improve the bottom line and the quality of life. Dick Mahoney, the former CEO and chairman of Monsanto, set aggressive goals for his company. Six years after starting in 1988, Monsanto reduced its toxic air emissions by 90 percent. Today, under the leadership of CEO and chairman Robert B. Shapiro, Monsanto is aggressively pursuing sustainable enterprise to gain a competitive advantage.

Many people are beginning to see the atmosphere, trees, and the ocean's plankton as tied directly to our own lungs and general well-being. The following is an example where citizen groups and individuals are using their "purchasing power" to lower the ceiling on allowable pollution into the atmosphere.

As part of a national program to control "acid rain" under the Clean Air Act Amendments of 1990, the U.S. Environmental Protection Agency and the Chicago Board of Trade have been auctioning "sulfur dioxide (SO_2) allowances" each spring since 1992. An allowance is a certificate that bestows to the bearer the right to pollute the atmosphere with a ton of gas. A purpose of the national SO_2 trading program is to lower the compliance costs of electric utilities seeking to control emissions that cause acid rain. A purpose of the annual EPA auctions is to ensure market fluidity and access to certificates by new entrants or startup companies.

About 3 percent of the total number of SO_2 allowances are made accessible to the general public and others through the annual EPA auction held each year. In 1997, 150,000 allowances were available in the "spot" auction for use during that year. Additionally, 125,000 allowances will be sold in the seven-year advance auction, usable for compliance beginning in 2004. And another 25,000 allowances were sold in the six-year advance auction, usable for compliance beginning in 2003. While economists predicted allowances might trade at just less than the cost of scrubbing ($600 to $800 per ton removed), the clearing price at the 1996 spot auction was $66.05, with a weighted average winning price of $68.14. In 1997, the value of SO_2 allowances was $106.75 per ton. Many environmental, school, and civic groups are buying allowances for the express purpose of "retiring" the pollution. By purchasing the certificate and never using it, they reduce the total amount of allowable SO_2 pollution.

These examples illustrate how people and organizations are recognizing that air is a vital flow supporting human enterprise and society within the broader life-sustaining ecology of the planet. In the case of global insurance companies and banks, they see degradation and modification of the atmosphere as subtracting from their bottom line. Business ecology goes beyond minimizing or diffusing the impact on air quality from economic activity. Instead, it recognizes that air is a basic, life-sustaining flow contributing to our economy. In this context, sustainable businesses and organizations will profit by developing products and services that actually improve air quality while creating value for stakeholders.

WE ARE WHAT WE EAT

Food is important to the business organism for a number of reasons: (1) a company's habitat includes space where employees and customers eat and interact socially; (2) the quality of food is directly related to employee health, well-being, and productivity; (3) dining experiences commonly build community within and among companies (e.g., business luncheons); and (4) food can renew our connection with the earth's ecology.

A company's habitat includes the space where employees and customers eat and interact socially. Food can be an important factor affecting your business's habitat, the quality of the work environment, and how it relates to the community. For example, are there decent eating establishments within walking distance for entertaining clients or conducting business meetings? Will a new building include a cafeteria and outside dining area for its employees? How will your company affect the economic viability of nearby food-related businesses? These considerations can affect how well you attract and retain good employees and customers, and how your company relates to its surrounding community. For example, decisions on food-related services for business events and regular patronage by employees can be vitally important to a community's economy. Eating and travel patterns during meal times, depending on the size of a company, can also affect traffic congestion and air pollution, particularly in the summer when urban ozone can be a problem. Walking, ordering in, and carpooling are effective strategies for reducing these problems. And finally, how employees conduct themselves in eating establishments projects your company's image to the community.

Because food sustains the life force that sustains your company, its no surprise that the quality of food consumed by your employees directly relates to your organization's well-being and productivity. While it is impossible to control how employees eat, convenient access to healthy, nutritional foods is good business. In general, employees who eat well, have *time* to eat, and exercise regularly (e.g., in a nearby or company gymnasium) are less prone to sickness and more alert and productive. Health insurance companies, for example, are aggressively promoting healthier lifestyles as a practical means of improving their bottom line though reduced healthcare payouts.

In our race against the clock, **we have forgotten the important social and spiritual dimensions of eating that are integral to being human.** It is not unusual in many fast-paced business settings for employees to "wolf down" and even skip meals. None of us are designed to behave as if we are parts of a perpetual motion machine. Employee dining experiences can provide time to process ideas, plan, and focus energy. They can also enable shared rituals that build community, both within and among companies.

Alice Waters, founder and proprietor of the Chez Panisse Restaurant & Cafe in Berkeley, California, is widely recognized for her leadership in the American movement toward healthy eating and for her support of community-based, organic agriculture. Waters believes that our relationship with food reveals much about our culture, and that food is a powerful means for building community-based values. The following is an excerpt from a letter she wrote to President Bill Clinton in December 1996:

> I may sound simplistic, but I continue to believe that the very best way to bring people together is by changing the role food plays in our national life. There is a growing consensus that many of our social and political problems have arisen because we are alienated from meaningful participation in the everyday act of feeding ourselves. The way in which we produce, prepare, and eat food expresses the bedrock values on which our public and private lives are built.
>
> In just two generations, the number of farmers has declined so much that very few Americans know anything about the people who grow their food. Even fewer have ever met a farmer, or know what it means to take care of the land. In the same span of time, a majority of American families have given up sitting down and eating together. How can we teach basic human values—such as courtesy, civility, honesty, and generosity—without that daily demonstration to our children that actions have consequences, that survival requires cooperation, and that people and nature are interdependent? These are precisely the lessons that are instilled in an elemental way by the family meal. It is also clear that they are not instilled by a reliance on processed and fast food, a way of eating that teaches us that we can fill our bodies with cheap, impersonal quick fixes and that eating is little more than refueling, devoid of any seasonal, agricultural, or social context.

Since 1971, Chez Panisse has consistently provided fresh, organic local foods in an atmosphere that exudes community. Chez Panisse and Alice Waters have received countless awards, including the International Women's Forum, The Woman Who Made a Difference Honoree (1987), Wine and Food Achievement (1989), Best Chef in America and Best Restaurant in America (1992). Referring to Waters as a "food revolutionary," Marian Burros of the *New York Times* had this to say in August 1996:

> Alice Louise Waters, one of four daughters born in Chatham, N.J., is no longer just a restaurateur. Chez Panisse, which she opened just to entertain her friends, has become a shrine to American cooking and a mecca of the culinary world.
>
> Now, Ms. Waters is on a mission: she believes that if all Americans had access to fresh, organic food, if all Americans worried as much about the environment as they do about potato chips, the world would be a better place.

Alice Waters has created a dynamic web of relationships that includes producers, suppliers, customers, business and community leaders, and employees. Chez Panisse is a wonderful illustration of the power of "putting community back in business." This community-based enterprise continues to challenge our perception of food and the important linkages we have with each other, the purveyors of food, and the Earth itself.

The Dan'l Webster Inn in Cape Cod, Massachusetts, is another example of business ecology in action in the food sector. The Inn uses an advanced water filtration system and a hydroponic/aquaculture greenhouse to provide high-quality water, fish, salad greens, flowers, and herbs to its dining guests. This closed-loop food system ensures that these fresh products are available and free of pollutants and pesticides. Since many of the products and services are produced, processed, and sold on the premises, this business captures significant benefits from value creation and reduced marketing, sales, and distribution costs. Strategic positioning in the ecotourism and health food markets ensures loyal repeat customers and generates positive cash flow.

Food, one of the original forms of currency, will become increasingly valued in a sustainable, ecological economy as we rediscover that our ability to produce nutritious, high-quality food is tied to a healthy environment and sustainable agriculture. Further, as pointed out by Alice Waters, healthy foods and eating patterns strengthen our businesses and organizations and society by reinforcing community-based values, such as interdependence, trust, and cooperation.

DISCOVERING THE LOST ECONOMY

A common business ecology thread weaves through the following case studies gleaned from David Gravely of Louisville, Kentucky, a uniquely forward-thinking and entrepreneurial individual. For several years, he and one of the authors carpooled from Louisville to Frankfort, Kentucky. In the course of this almost daily commute, Gravely, who specialized in imagining ways that waste materials and by-products could be converted profitably into useful products and services, shared his thinking, which was forty years ahead of its time. Why is it still relevant today? Because its efficiency and closed-loop thinking mirrors natural systems and creates common sense profits.

Cigarettes and Whiskey

In the 1950s, at Louisville's large tobacco companies, one of the items used in packaging cigarettes was a foil-backed paper. These wrappers ensured that each pack of cigarettes would remain fresh until opened. The raw material for this wrapping was shipped to the cigarette packing facility in huge rolled sheets. Later, this rolled foil would feed through paper shears, or punches, to cut the small rectangular pieces of foil actually used in each pack of cigarettes. As a consequence of the design of the machinery, small amounts of foil were left over from around each piece. Due to large volumes of throughput at the cigarette companies, this small amount of scrap would accumulate quickly, being packed into many old 55-gallon drums. This waste was a problem because the foil and paper were bonded so tightly together; it would not burn easily because of the foil, nor melt easily because of the paper. So, it was sent to the landfill for a nominal fee.

Dave Gravely saw an economic opportunity, since he knew that the key to separating the foil and paper was roasting, not open burning or crucible melting. Using scrap materials and a home shop, he assembled an inclined rotating cylindrical furnace of his own design that had two key features: an oxygen-deprived combustion zone, and a tumbling motion to ensure good heat transfer from the outside heat to the charge of scrap foil within. In operation, the furnace resembled an industrial-sized version of a rock-tumbler for polishing stones. The tumbling motion also helped separate the carbonized paper and aluminum foil. The operation produced two useful and salable products: lamp black for inks, dyes, and typewriter ribbons, and aluminum ingots.

Once the concept was proven to his satisfaction, Gravely offered to take the waste from the cigarette companies for a price just slightly cheaper than the companies were already paying. When it became clear that a cheap, steady, reliable source of fuel would be needed to sustain operations, he chose to fuel it with waste wood scraps from construction sites and other sources. Rather than calling up the local natural gas company for a connection, Dave applied his technique, insights, and vision to the problem of fuel and created three additional income streams for his growing company.

Rather than paying for fuel, he looked for ways in which the fuel problem could be solved profitably, preferably by someone who was paying to dispose of something with heat value. The vision for a profitable solution? Another Louisville industry, whiskey-making. By understanding the process and life-cycle of whiskey production, Gravely found another niche opportunity. Good whiskeys are made by aging distilled spirits in white oak barrels for five to seven years or more. Before filling, each new barrel is scorched on its inside to carbonize the surface, creating the source of the caramel color found in fine whiskeys. The top-end distillers use the barrels only once; then they are often sold to mid-range distillers who use the barrels several more times. Eventually, the used barrels are sold yet again to the low-end distillers where they may be used several more times. By this point the barrels are quite old, the carbonized inner surfaces have grown chemically weary, and in the 1950s, only "moonshiners" saw any value in the old barrels—and Dave Gravely, who needed a fuel source. By applying his oxygen-deprived combustion technique to the outer combustion chamber, a fine white oak charcoal was produced, bagged, and sold as fuel for outdoor barbecues. One additional benefit was realized from what seemed to be a problem. As the white oak staves burned in an oxygen-poor environment, they produced a tarry substance that threatened to reduce the heat transfer to the inner chamber of the furnace. A chemist friend analyzed the "gooky" substance in a laboratory and discovered the presence of an enzyme that is critical in making a good meat tenderizer. So with further refining and extraction, an additional revenue stream was produced by selling this product to a food company. What started only as an idea was now producing six separate streams of income and efficiently creating useful cycles of reusable materials.

Alcohol and Sulfuric Acid

Years later, Dave Gravely helped dispose of alcohols that had become tainted and consequently unfit for use in pharmaceuticals. Several large Midwestern pharmaceutical companies had this problem and were paying to have alcohol wastes removed from their facilities. Gravely once again set about efficiently reusing flows of waste materials with the beneficial side-effect of creating multiple income streams. He located an efficient alcohol combuster of his own design on a property adjacent to a company that used plenty of heat for steam cleaning and reconditioning 55-gallon barrels. The source of this heat energy was the local utility company. Gravely offered his barrel-reconditioning neighbor heat energy at just below the market rate. The mechanics were simple: run an insulated pipe from one facility, through a chain-link fence, to the next. Twin streams of income were produced—one from the pharmaceutical company and one from the barrel reconditioner.

With growing environmental concerns over their other waste streams, the drug companies were looking to consolidate waste vendors. This would mean testing existing vendors to see if they were robust enough to handle larger volumes and other wastes, such as sulfuric acid. Both of these tests were put to Gravely's small facility, originally designed to handle alcohol only. Gravely knew that he would need to say "yes" to an extra large shipment of sulfuric acid to stay in the disposal game, and arranged for delivery of some large second-hand tanks from local salvage yards. When the sulfuric acid tankers arrived, they had a much greater volume than he was prepared to handle; the demurrage charges on the tankers would mount quickly if he failed to find a solution.

Gravely once again seized an opportunity from a problem. He recalled that his neighbor, the drum reconditioner, had mentioned that he had acquired a shipment of drums that were proving very difficult to clean. The drums previously held an epoxy-like substance that left behind a very problematic coating that was resisting all efforts to remove it, and it looked like a financial loss to the drum-cleaner. Gravely inquired of his neighbor whether he would consider subcontracting the cleaning job to him, at a modest cost. Relieved, the man agreed, and soon thousands of gallons of much-needed storage capacity were coming through the gate in the chain link fence. And once filled with sulfuric acid, the troublesome coating eventually yielded. Gravely once again had found twin income streams—one from the pharmaceutical company and one from the barrel reconditioner.

But how to use the sulfuric acid? With the aid of a chemist friend, Gravely created equipment to purify the sulfuric acid and to bring it to the specifications of battery acid. The next stop for this acid was a major battery manufacturer and a major supply house for after-market automotive parts. Together, these sources produced the greatest amount of income from the entire sulfuric acid venture. Most importantly, a source of steady supply from the pharmaceutical company was not disrupted because the small company had passed the tests.

Dave Gravely's ability to see more than one use for materials and resources is essential to putting business ecology into action—such closed-loop thinking is both environmentally and economically beneficial.

CLOSING THE LOOP: HOW SYSTEMS THINKING CAN HELP YOUR BOTTOM LINE

Business ecology is a lens for systemically seeing your company's flows. It strengthens your company's viability by helping you:

1. See and improve your company's metabolism
2. "Close the loop" with other companies
3. Develop new products and services from residual flows
4. Create multiple profit cycles

Combined with the knowledge and insight to know what to do, and the creativity and boldness to make it happen, business ecology can make resource recovery and reuse a more viable, lucrative option. For business ecosystems, such as the Danish Kalundborg facility or even Dave Gravely's small companies, to thrive and prosper, they or their organizing agents must understand the ecological flows of individual companies and their neighbors. When such interconnections are mindfully made, everybody wins. This leads to profitability, healthy stakeholder relations, and a cleaner environment.

More recently, systemic analysis of companies and their metabolic flows have become more commonplace. Good examples are the emergence of energy audits and broader-based ecological audits (eco-audits), and the development in 1995 of ISO 14000 Environmental Management Standards (see Chapters 1 and 2). **In the context of business ecology, the ultimate goal of eco-audits is to improve simultaneously metabolic efficiency, environmental performance, and profitibility. Eco-audits simply extend the systems-based accounting to other flows such as water, materials, employees, and information.** The following examples illustrate how businesses are applying ecological thinking to improve their bottom line.

Ray C. Anderson, chairman and CEO of Interface, Inc. in Atlanta, Georgia, is aggressively moving his one billion dollar carpet tile and interior furnishings company toward a sustainable future. Interface has 5,300 employees who manufacture products on four continents and sell them in 110 countries. Here is how Anderson sees the situation:

> The world is in real trouble. I'm convinced that my company and every company on earth is doing damage that's unconscionable. The only institution on earth that is capable of making a difference is business and industry.

Already, Interface has saved about $13 million by reducing waste. In addition, Anderson is working with The Natural Step—U.S.A. (see Chapter 7) to help his employees learn basic natural principles or "system conditions" that will govern the future of their company and society as a whole. Here is one of Anderson's challenges to his employees: "We have to learn how to harvest carpet that's already been made. Billions and billions of yards of carpet are out there. We need to recycle the carpet to salvage the nylon."

Here is an example from the American textile industry that illustrates how economic, energy, and environmental performance improvements can work together. The textile industry uses more than 100 billion gallons (378.5 billion liters) of water each year. Most of this water is used for preparation and dyeing processes. Water is pumped into a dyeing machine, where fabric is placed in the bath and saturated with water. Chemical auxiliaries (e.g., wetting agents, pH control agents, leveling agents, and chelating agents) and dye are added to the bath water. Temperatures are raised at a rate of about 2°C (36°F) per minute until dyeing temperature is reached, and then held constant until dyeing is complete. Afterward, the dye-bath is emptied, then refilled to begin the process again.

Large quantities of energy, water, and useful chemicals are lost when the dye-bath is discharged. To address this issue, the Georgia Hazardous Waste Management Authority, the Georgia Institute of Technology, Shaw Industries and others are developing and testing an automated system that analyzes spent dye-bath and reconstitutes the bath to meet dyeing requirements. The project is supported though the National Industrial Competitiveness through Energy, Environment, and Economics (NICE[3]) grant program at the U.S. Department of Energy.

The NICE[3] program estimates that dye-bath reuse will reuse 42 percent of water, 60 percent of auxiliary chemicals, and 6 percent of dyes. With an initial investment of $832,741 (including an industry share of $396,641), the dye-bath reuse technology could save 3.6 trillion BTUs (3.8 quadrillion joules) per year industry-wide. These savings are derived from three sources: (1) the reduction in direct thermal and electrical energy to heat dye-baths; (2) the elimination of energy to produce dyes, auxiliary chemicals, and water; and (3) reduction in energy associated with treatment of wastewater. Thirty-six million pounds of auxiliary chemicals could be reused rather than dispersed into the environment. Shaw Industries saves about $1.6 million per year in carpet production costs per plant.

In Ohio, the AAP St. Mary's Company integrated an innovative aluminum chip recycling process into its existing wheel production facility instead of outsourcing to a recycling firm. By "closing the loop" with this new, more efficient technology, pollution related to cutting oils, airborne fumes, and dust has been prevented or sharply reduced while production, energy efficiency, and quality control have improved dramatically. The bottom-line benefit: AAP St. Mary's Company saves about $1.92 million annually in operating and maintenance costs with an investment payback of 1.6 years.

What about an eco-efficient, closed-loop system that creates savings for the company and passes savings along to homeowners? In North Carolina, Enertia® Building Systems, Inc. (EBS) manufactures elegant, off-the-grid, solar-geothermal homes that act as house-scale heat pumps, using double-wall design and solar and geothermal resources to provide year-round comfort for homeowners without furnaces or air conditioning units. In addition to the significant cost-savings to homeowners, these homes dramatically reduce the environmental footprint of home-building through full life-cycle, ecological accounting and management. To mitigate the obvious impacts from harvesting trees for new construction, EBS, Inc., for example, uses local, sustainably harvested trees and plans to manufacture their homes within regional business ecosystems close to new home sites. Michael Sykes, inventor and president of EBS, Inc., describes the energy used to make one of their homes in the context of more traditional homes:

> The Enertia® house is made of wood, the only renewable natural material, with a minimum of energy-consuming processing. Brick, which has to be baked in ovens, requires many times more energy before you can even use it. Even particle boards, which are made of wood waste, require millions of BTUs of energy when they are glued, pressed and cured. The total energy *to make and use* an Enertia® House for a lifetime, is less than 10 percent of the energy just to make materials for a conventional fuel-using home.

Another example of ecological enterprise, the Mason Dixon Farm, Inc. in Gettysburg, Pennsylvania, converts animal waste to methane and sludge with an anaerobic reactor. The methane provides for all of the farm's electricity needs, with about $90,000 per year of excess energy sold to the local utility. The sludge is applied to the fields directly as fertilizer, offsetting the need to purchase synthetic fertilizers. In this way, pollution is also prevented.

Each of these companies has found that closed-loop thinking—a cornerstone of business ecology—has revealed creative solutions, improved their company's profitability, and passed benefits along to their customers and communities as well.

Decision Dynamics, based in Maine, provides decision-support for ecological assessments. Founder Greg Norris has developed a unique software system for identifying and evaluating profitable investments related to pollution prevention and improving a company's metabolic efficiency. He has tested the software with Caterpillar, Inc., a leading U.S. manufacturer of earth-moving equipment. PTLaser™ software integrates process modeling with life-cycle cost analysis. For instance, it provides manufacturers with an inventory of their energy and raw material inputs, and their waste and pollution as product outputs. It also provides a "total cost assessment," which breaks out costs by activity or process, and by flow of material, energy, or waste. In addition, users of the software can rapidly evaluate the cost and environmental implications of a particular process or product modification and explore new opportunities by calculating the sensitivity of present value costs to any relevant variables.

Many business people consider this closed-loop "eco-efficiency" to be a natural extension of total quality management. And like other performance improvements, it has real bottom-line benefits. Business ecology takes it a step further by providing a systemic lens for seeing how to improve your organization's ecological efficiency and its viability. Figure 4.1 is such a lens, a systemic tool for assessing simultaneously your organization's life-sustaining flows. Products and services, water, food, materials, energy, air, people and other organisms, information and ideas, and money are plotted to give a composite picture of organizational viability. The nine plots are equally spaced along the arcs of the circle to convey an ideal sense of balance among the several flows. The points on each of the radii inside the circle are a measure of the viability of each of these flows. Points plotted closer to the circumference are more viable than those closer to the center.

Developing a sustainable business or organization requires that managers and leaders become more aware of life-sustaining flows, much like an organism is aware of its food, water, and shelter. The viability of each

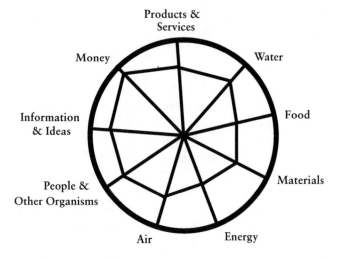

Figure 4.1   How Viable Are Your Organization's Flows?
© 1998 Business Ecology Associates

flow shown in Figure 4.1 can be determined by a variety of quantitative and qualitative factors, including:

- Strength and sustainability of gradients that induce flow (i.e., how well does your organization or community attract and secure financial investments?)
- Distance from source(s) (i.e., how far is your business from its electric power source?)
- Sustainability of source(s), (i.e., how dependent is your company on petroleum and other forms of nonrenewable energy?)
- Sustainability/reliability of flow system(s), (i.e., how well-maintained are your community's water and sewage systems?)
- Efficiency/life-cycle costs, (i.e., how energy-, water-, and material-intensive are your products and services?)
- Cyclicity—the degree to which flows are closed-looped

A systemic profile for *each* flow, such as energy, food, or water, similar to Figure 4.1 can also be plotted with respect to the above and other flow attributes. Why are the above flow factors critical to creating and understanding sustainable enterprise? Electricity, a vital energy flow sustaining most organizations, provides an illustrative example.

A utility delivers electricity to paying customers via a complex flow system, or power grid, that includes a power plant, high-voltage lines, transformers, utility poles, and other facilities to ensure reliable service. The "food web" supporting this metabolic pathway varies, but generally includes a combination of local, regional, and global resource flows. For example, utilities may use energy such as coal, petroleum, natural gas, nuclear, and other sources, and materials, such as timber, copper, water, and steel, to generate, distribute, and sell electricity to customers over regional/local power grids. Coal burned for electricity generation, for instance, may be mined hundreds of miles from a power plant. Timber for utility poles may be harvested over a thousand miles away. In other instances, petroleum fields supplying a power plant may be halfway around the world. This means even more energy, materials, and other natural resources are used to discover, develop, and transport natural resources that are consumed directly by the power system.

While estimates vary on just how much fossil fuel remains, the fact that deposits of fossil fuels are finite is certain. Our fossil-fuel-based economy, including the way we produce, distribute, and use electricity, is not sustainable in the long run. Some industry estimates, for instance, indicate that the world oil reserves will be depleted by the year 2050 at current consumption rates. While coal reserves are significantly greater, burning coal leads to acid rain, climate change, acid-mine drainage, and other significant environmental problems. Intractable problems with nuclear power, such as long-lived radioactive waste, public distrust, and life-cycle costs of fuel production, seriously hamper it as an option.

Power stations are frequently remote from customers for several reasons, including: (1) to allow atmospheric diffusion of plant emissions in less populated areas; (2) plants tend to be sited in areas where there is little public or political opposition; (3) ready access to transportation, water, fuel, and other resources. This remoteness often discourages efficient recycling of residual flows, such as steam, heated water, and emission control by-products. Customers are also less inclined to connect their electricity consumption with pollution problems, such as acid rain, particulate matter, thermal water pollution, heavy metals, and climate change.

Finally, while the benefits of shifting toward more sustainable electricity systems are obvious, there are significant economic barriers. Since the total cost of providing electricity is not reflected in utility rates, there is little incentive in the marketplace to shift to more sustainable technologies, such as those discussed in the section called "Powerful Insights" above. Prices, subsidies, and regulations—all human constructs—frequently create economic gradients and flows that defy common sense and ecological principles. As long as prices do not reflect the total life-cycle costs of electricity, the U.S. economy will continue to be dependent on ecologically and economically inefficient power systems.

Clearly, new opportunities for economic efficiency and pollution prevention are realized by adopting a "whole systems" perspective rather than the more traditional accounting methods that focus exclusively on cash flow and the bottom line. The insights gained from a whole systems look into a company's metabolism are crucial to developing healthy business ecosystems, which are discussed in the next chapter.

5

Business Ecosystems

Perhaps the most poignant image of our time is that of earth
as seen by the space voyagers: a blue sphere, shimmering with
life and light, alone and unique in the cosmos. From this perspective, the maps of geopolitics vanish, and the underlying
interconnectiveness of all components of this extraordinary
living system—animal, plant, water, land and atmosphere—
becomes strikingly evident.

R. Benedict

We have much to learn from living systems. Our future, the future of
humankind, will depend on our ability to reconnect to the natural world,
and rediscover its life-supporting, organizational intelligence. Solutions exist
all around us. We can see them when we have the humility to remove our
self-imposed, human-centered blinders. With this new vision, we can learn
and apply nature's organizing elegance to create a sustainable future, one
that balances our needs with the rest of the living world.

In Chapter 1, a business ecosystem was introduced as a dynamic
community of companies and other "organisms" that echoes natural systems. Cyclical and mutually beneficial relationships develop within and
among firms, institutions, and communities to share vital resource and
commercial flows such as money, energy, materials, information, water,
food, and people. In such a system, surplus flow from one entity is a
valuable resource for another. This chapter will: (1) examine various
business ecosystems from around the world; (2) explore what we can
learn from agro-ecosystems; and (3) discuss the challenges and opportunities related to healthy business ecosystem development.

BUSINESS ECOSYSTEMS AROUND THE WORLD

Following are brief profiles of several business ecosystems at various stages of development:

- Kalundborg Region, Denmark
- Burlington County, New Jersey, U.S.A.
- Cape Charles, Virginia, U.S.A.
- Chattanooga, Tennessee, U.S.A.
- Trenton, New Jersey, U.S.A.
- Brownsville, Texas, U.S.A.
- Matmoros, Mexico
- Baltimore, Maryland, U.S.A.
- Skagit County, Washington, U.S.A.

Each of these business ecosystems is uniquely adapted to local needs and resources. In the context of business ecology, companies are "organisms" participating in "business ecosystems."

The term "eco-industrial park" or "EIP" is used frequently within these profiles, as well as within the field of industrial ecology. The authors consider this term somewhat limiting, and even inaccurate, in that it implies a specific tract of land or parcel. The examples described here are actually systems of organizations that exchange resource flows, often located at different sites, which include various "system levels," such as company, community, and regional economy. Second, the term "business" is more encompassing than "industrial," since this latter term generally refers to energy- and material-intensive manufacturing and processing, and not commercial or service enterprises. Thus, business ecosystem is preferred to eco-industrial park. An eco-industrial park, or "industrial ecosystem," is one type of business ecosystem.

Kalundborg Region, Denmark

The Kalundborg, Denmark, business community is an excellent example of a business ecosystem. Kalundborg is a mecca for industrial ecology. Many research, business, and community groups in North America, Europe, Japan, and elsewhere have been studying this business complex in the hopes of replicating it. Here's an in-depth discussion of the Kalundborg Region, based on information provided by The Symbiosis Institute in Kalundborg, Denmark. Organized in the 1990s under the auspices of the

Industrial Development Council in cooperation with enterprises participating in the Kalundborg complex, the Symbiosis Institute gathers, records, and propagates information on the Kalundborg Region and similar business ecosystems. The key aspects of the Institute's mission are education and building strategic alliances to encourage similar business ecosystems.

Since the 1970s, the Kalundborg Region of Denmark has developed a remarkable pattern of commercial and resource exchange that in many ways resembles a natural ecosystem. It was not until 1990, however, that Kalundborg's "industrial symbiosis" gained international attention. Here's how the Symbiosis Institute defines industrial symbiosis:

> Symbiosis is the living together of two dissimilar organisms in any of various mutually beneficial relationships. Here, the expression is used about an industrial cooperation between several companies which trade by-products with each other.

Networking among participating companies, based on common values that include sound commercial principles, is vital to developing and maintaining the collaborative exchange of by-products:

> The industrial symbiosis in the Kalundborg district is built up as a network cooperation between four industrial enterprises in the town and municipality of Kalundborg. In this symbiosis, the four enterprises: The Asnæs Power Station, the plasterboard manufacturer, GYPROC, the pharmaceutical and biotechnology company, Novo Nordisk, and STATOIL refinery trade by-products because waste of each is a valuable raw material to one or more of the others. The result is a reduction of both resource consumption and environmental impacts.
>
> These four business partners also gain financially from the cooperation because all contracts within the symbiosis are based on sound commercial principles.

Figure 5.1 provides an overview of the Kalundborg business ecosystem. As shown in Figure 5.1, the Asnæs Power Station is the central, anchoring organism in the business ecosystem, producing heat for the town of Kalundborg, process steam for the STATOIL Refinery and Novo Nordisk, and coolant water to an aquaculture business. Five thousand households in Kalundborg receive district heat from Asnæs Power Station, replacing the burning of approximately 3,500 small oil-fired units. Twenty percent of STATOIL's process steam, used in heating oil tanks and pipelines and other purposes, comes from the power plant. Novo Nordisk uses the

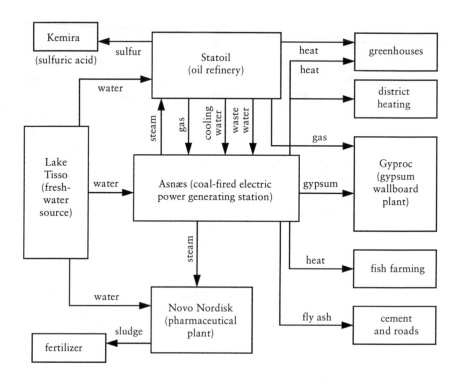

Figure 5.1   An Overview of the Kalundborg, Denmark, Business Ecosystem
Source: Novo Nordisk

steam for heating and sterilization purposes. A local fish farm, producing about 200 tons of trout a year, uses cooling water from the power plant to create productive growing conditions.

Similar to other coal-fired power plants around the world, the Asnæs Power Plant "scrubs" sulfur from its flue gas to comply with air quality standards. The plant's flue gas desulfurization system removes sulfur dioxide from the flue gas, producing cleaner emissions and 80,000 tons of gypsum a year. The gypsum, which is purer than natural gypsum, is sold to GYPROC, where it is well-suited for the production of plasterboard. This source of gypsum meets about two thirds of GYPROC's needs. Another by-product of Asnæs Power Station is fly ash. About 135,000 tons of fly ash is used in cement and concrete production locally.

GYPROC and Asnæs, in turn, use surplus refinery gas from the STAT-OIL refinery. This gas replaces 90 to 95 percent of oil consumption used by GYPROC to dry wet gypsum as part of the plasterboard production process. The remainder of drying heat is provided by butane gas. About 30,000 tons of coal per year (about 2 percent of its total fuel consumption) is substituted by refinery gas at Asnæs Power Station.

STATOIL also cleans its emissions to comply with air standards. Its desulfurization plant removes sulfur dioxide from its flue gas to produce about 3,000 tons of pure sulfur annually. This is used by Kemira, a small local company, to make sulfuric acid.

Novo Nordisk ferments raw materials such as potato flour and corn starch to make industrial enzymes. About 4,000 to 5,000 cubic meters of biomass by-product is created each day. With lime added to neutralize pathogens, the nitrogen and phosphorus-rich biomass is transported and sold to farmers as fertilizer, reducing their need for commercial fertilizer. Novo Nordisk also produces insulin for the treatment of diabetics. Surplus yeast, a by-product of insulin production, is used as animal fed.

The Symbiosis Institute suggests that the following conditions are necessary to encourage industrial symbiosis:

The enterprises must function together:

It must be possible for waste products from one enterprise to be used by another. In Kalundborg, half of the enterprises produce energy, while all of them use it. Industrial symbiosis is practiced elsewhere as well, but rarely between more than two enterprises.

The enterprises must be situated near each other:

For symbiosis to work, the enterprises must be reasonably near each other. Long pipelines are costly and the greater the length, the greater the energy losses. Experience from Kalundborg shows that distance is most important when it's energy that is being exchanged between enterprises. Distance is of less importance in the case of other by-products.

Openness between the enterprises:

With its relatively small size and isolated situation, Kalundborg has provided fertile soil for an industrial symbiosis because the enterprises working there all know each other and have developed a relationship characterized by openness, communication, and mutual trust.

Burlington County, New Jersey, U.S.A.

Another example of a business ecosystem is the Burlington County Resource Recovery Facility in New Jersey, U.S.A. A state recycling law was the impetus for the creation of this recycling processing complex and a complementary research center called the EcoComplex. The first tenant of the business ecosystem, which is located next to the county landfill, was a materials recovery facility (MRF). It receives 150 tons per day of paper and mixed recyclables collected from county households. Next, a large-scale biosolids co-composting facility was added to capture additional materials destined for the landfill. The availability of residual wood from local construction projects prompted the purchasing and installation of wood shredding and screening equipment; wood chips produced by this equipment are fed into the composting facility.

Here are some of the county's plans for the future. It plans to build manual and mechanical sorting systems to ensure that the source wood for the composting is "clean." A pipe foundry intends to use 350,000 scrap tires annually as fuel. A $2 million tire processing facility would be located near the MRF to solve both the foundry's fuel needs and a long-standing discarded tire management problem in Burlington and other counties. Some of this ground rubber may be sold to an asphalt plant to be built nearby. This asphalt plant is interested in buying methane from the landfill and using the ground rubber and recovered asphalt shingles in its production.

Harry Janes, a Rutgers professor, is directing the development of a $6 million EcoComplex to help spawn environmental technologies that would complement the Resource Recovery Facility. Examples include a methane-powered greenhouse and the productive use of mined landfill material. Here is Janes' vision for the EcoComplex, which will be:

> a magnet that attracts new opportunities and funds, bringing together teams to address subjects such as plastics recycling, coproduct use, bioremediation, detoxification of waste streams, treatment of biosolids, development of good agricultural practices for growing high-value nonfood crops, and innovative treatment of food processing waste streams.

Cape Charles, Virginia, U.S.A.

The Port of Cape Charles Sustainable Technologies Industrial Park in Northampton County, Virginia, is attempting to make industrial ecology work in the United States. It is one of four U.S. demonstration projects initiated in 1995 by the President's Council of Sustainable Development, a consortium of business, government, environment, and community leaders. The three others include Baltimore, Maryland; Chattanooga, Tennessee; and Brownsville, Texas/Matamoros, Mexico.

Cape Charles, Virginia and the surrounding areas of the Eastern Shore of the Chesapeake Bay see sustainable redevelopment as vital to their future. Cape Charles began as a town built around a railroad terminus and ferry that linked the Eastern Shore with Norfolk, Virginia. However, the completion of the Bay Bridge Tunnel redirected traffic (and commerce) around Cape Charles. Today, the local economy primarily consists of agriculture, a concrete company (which helped build the bridge/tunnel system), and an underutilized port. Abandoned stores, restaurants, shops, businesses, and homes dominate this tranquil village.

Although economically depressed, the citizens of Cape Charles describe where they live as

> a place rich in natural and cultural assets—beaches, islands, marshes, woodlands, tidal creeks, fish and shellfish, birds and wildlife, open land, clean water, historic villages and farms, skilled people, hardworking people, people of proud heritage.

Today, building on that heritage, Cape Charles is preparing for a renaissance, according to local leaders such as Bruce Evans, innkeeper, Timothy Hayes, sustainable development director, and Keith Bull, economic development director. Since the spring of 1995, citizens, businesses, government, and nonprofit organizations have been working together to make their shared vision for Cape Charles a reality. The vision focuses on attracting sustainable enterprise to the town and surrounding areas that may be linked to existing economic, social, and environmental resources already available in the region.

In October 1996, the Port of Cape Charles Sustainable Technologies Industrial Park "broke ground," marking the beginning of the redevelopment project. Solar Building Systems, Inc., a company that assembles solar cells into panels, is the first tenant in the park. In addition to the attractive nature of the park itself, additional economic incentives for the firm included a National Enterprise Community Grant and Virginia's Photovoltaic Manufacturing Incentive Grant Program. Phillip Leach, managing director of the company, believes the park will incorporate both new environmental technology firms as well as more traditional companies, such as those producing roofing and building materials.

Maintaining an excellent business environment, a pristine natural environment, and an exceptional living environment are key goals for the Cape Charles project. Enertia Building Systems, the solar geothermal home builder profiled in Chapter 2, has expressed an interest in locating one of their plants in Cape Charles. Cape Charles is also working to attract aquaculture firms, resource recovery industries for agricultural materials, hydroponic gardens, organic farms, and ecotourism companies.

Chattanooga, Tennessee, U.S.A.

Recognizing that "development as usual" is not working, a number of communities across the U.S. and abroad have initiated sustainable development projects. Chattanooga, Tennessee, is a good example of a community that is embracing sustainable development opportunities. It has already undergone a rebirth in community spirit and economic revitalization, and is quickly developing a reputation as one of the most environmentally friendly cities in the U.S. and possibly the world. This profile constitutes quite a change, as this mid-size city—which Chattanooga Council member David Crockett calls "the buckle of the Bible belt"—was once considered one of the most polluted communities in America. Chattanooga's restoration of the Tennessee River has become a powerful symbol for the city's rebirth since the 1970s. Building on this restorative theme, the Chattanooga Eco-Industrial Park Initiative includes four separate components: a brownfield reclamation park; a mixed-use site; an eco-industrial park; and an environmental technology complex.

A thirteen-acre site located in Chattanooga's older industrial area, formerly owned by Anchor Glass, has been targeted as the brownfield reclamation park. It is close to a low-income, single-family residential neighborhood and Chattanooga Creek, a federal Superfund site. In addition to environmental remediation, the strategy for this property and the adjoining neighborhood includes establishing training facilities for construction, neighborhood enhancements, and light manufacturing.

The Mixed-Use Greenfield Park/Lookout Valley Site, located on 775 acres five miles outside the city, is intended to attract ecologically sound manufacturing companies alongside housing and other employee amenities. Formerly the Volunteer Army Ammunition Plant, the 6,800-acre Eco-Industrial Park had been used for the production of TNT. With proper safeguards and liabilities taken into account, the plans call for two 300-to-400-acre industrial/business/warehouse distribution parks. A large 1,000-acre industrial site, a business incubator and technical school, and a National Environmental Test Center for developing and testing new technologies, are also planned.

In the South Central Business District, Gunter Pauli, a Belgian businessman and founder of the Zero Emissions Research Institute in Japan, is helping to develop a 100-acre Environmental Technology Complex. It is expected to house zero-emitting industries next to residential housing and incubator facilities.

Trenton, New Jersey, U.S.A.

Trenton, New Jersey, is home to a diverse, small, and declining population. The fortunes of this once-proud manufacturing mecca, which boasted "Trenton makes, the world takes," have declined. Many cities in the northeast U.S. have experienced decades of economic decline and job loss. Factories throughout the mid-Atlantic and Northeast regions have closed, many relocating elsewhere. In 1993, the Trenton Enterprise Initiative marked the beginning of a turn-around for Trenton. This community-wide planning process identified social, economic, and quality of life issues and strategies for addressing them. Trenton, once a renowned leader in manufacturing, is investing in sustainable development as a means for regaining jobs, revitalizing the community, and developing a competitive edge for the twenty-first century.

In August 1994, representatives from the New Jersey Department of Environmental Protection, Kean College, local industry, the U.S. Environmental Protection Agency (represented by author Joe Abe), nonprofit organizations, and the City of Trenton initiated a series of monthly meetings to define and design the components and operating principles of an eco-industrial park complex in or around Trenton. These on-going meetings, a follow-up to the Trenton Enterprise Initiative, became known as the Trenton Eco-Industrial Roundtable.

The mission of the Eco-Industrial Roundtable is to demonstrate a new paradigm for competitive business development that is based on environmental linkages and social responsibility. The Roundtable recognized the need to create a strategic plan to link industries and businesses to create enough inherent advantages so that companies will remain in, or will relocate to, Trenton and the surrounding region. Identified advantages include: lower costs of doing business through cheaper inputs; lower energy costs; less waste energy and materials; reduced shipping, regulatory and liability costs; a competitive edge resulting from these efficiencies; positive publicity for Trenton; and an image for participating businesses as being environmentally responsible, socially progressive, and innovative.

Some of the business opportunities identified through the Eco-Industrial Roundtable include aquaculture, hydroponics, and microbrewery enterprises. Power plants and district heating are in or near Trenton, providing a ready source of inexpensive heating for tanks and greenhouses. Spent grain from a microbrewery can be used to feed farm-raised fish. Greenhouses can house both aquaculture and hydroponic systems. Potential products from these systems, such as fresh fish, vegetables, herbs, flowers, and specialty plants, can be sold to restaurants, grocery stores, and florists in Trenton and the adjoining region.

Recycling of paper, cardboard, textiles, plastics, tires, metal, furniture, and building materials were also identified as good business opportunities for Trenton for a number of reasons. Population density, existing recycling programs, labor availability, access to rail, water, and highway, and proximity to major urban areas all make Trenton ideally suited for recycling and demanufacturing businesses. The Cornell Work and Environment Initiative was commissioned by Trenton in 1996 to develop a strategic plan based on information gathered from the Roundtable discussions. Trenton is actively seeking financing and recruiting businesses that fit their sustainable redevelopment vision.

Trigen-Trenton Energy Company, L.P., a company already situated in Trenton, provides electricity and district heating and cooling to the community, and is one of several businesses that may act as "anchor organisms" for future business ecosystem development. Don Liebowitz, Vice President of Trigen-Trenton, describes how his company can provide a "backbone of energy" for Trenton and similar communities pursuing sustainable redevelopment (see the section entitled "Powerful Insights" in Chapter 3).

Brownsville, Texas, U.S.A./Matamoros, Mexico

The Brownsville Eco-Industrial Park, located in the Brownsville, Texas, U.S.A./Matamoros, Mexico border region, promotes economic efficiency by facilitating exchanges of by-products and wastes among a community of manufacturing and service businesses. The project, cosponsored by the Brownsville Economic Development Council (BEDC) and the City of Brownsville, intends to curb environmental problems associated with rapid industrialization while creating a lucrative business environment. The project has a two-prong strategy: (1) develop a physical site for companies to co-locate to exchange by-products; and (2) develop a virtual, regional network of exchanges with existing companies in their present locations.

According to David Cobb of Bechtel Corporation, the principal consultant for the project, "Industrial symbiosis is where commercial development is headed." Cobb has developed a computer model that matches raw materials and fuels from some industries with waste and by-products from others to create a network of exchange like the Kalundborg, Denmark, business community. This model will help link existing businesses and identify "niches" to be filled by companies seeking to relocate to the region. Rick Luna, an official with BEDC, sees the ultimate goal of the network as "zero emissions."

Scenario planning has also been used to help chart different development paths. Each scenario represents a different set of prevailing assumptions. Projected case studies are being used to examine how existing port members, such as a refinery, a stone company, an asphalt company, and tank farms might interact with each other. Potential remote members also are being considered, including a textile plant, an auto parts manufacturer, a plastics recycler, a seafood processor and cold storage warehouse, a chemical plant, and a magnetic ballast manufacturer.

Baltimore, Maryland, U.S.A.

The Fairfield Ecological Industrial Park, located in the heart of Baltimore, Maryland, represents the most heavily industrialized project among the four demonstration projects initiated by the President's Council on Sustainable Development. Ed Cohen-Rosenthal and others from the Cornell Work and Environment Initiative, have been working with a consortium of business and community groups organized by the Baltimore Development Corporation. A pragmatic development strategy is being adopted, notes Cohen-Rosenthal:

> The primary goal on an EIP (Eco-Industrial Park) is to improve the economic performance of the participating companies while minimizing their combined environmental impact. . . . Components of this approach include new or retrofitted design of park infrastructure and plants; co-location of companies; joint pollution prevention and waste minimization strategies; ongoing energy/environmental auditing; ongoing research and development for materials substitution and replacement; energy efficiency assessment; and public and private sector

partnering. Through collaboration, this community of companies becomes an "industrial ecosystem." From an environmental perspective, the goal of an EIP is to demonstrate best environmental performance. It seeks higher performance with less risk.

Embedded with the above approach is a total quality environmental management (TQEM) philosophy, which calls for continuous improvement in the use of resources.

The cluster of industries at the Fairfield site predominantly are related to petroleum and organic chemicals. The Fairfield "carbon" economy includes oil company marketing sites, asphalt manufacturing and distributing, and divisions of multinational chemical companies making cleaning solutions, herbicides, and plating solutions. Companies at the Fairfield site include BP, Texaco, Mobil, Conoco, Shell, Sun Oil, Rone-Polenc, FMC, Vista Chemical, Clean America Corporation, Seaford Asphalt, and Colonial Pipeline.

According to a 1995 baseline study by the Cornell Work and Environment Initiative, smaller companies have sprouted up to support the needs of these large industries. These include trucking companies, rail and port services, environmental companies, box manufacturing, tire retreading, and materials handling machinery.

Skagit County, Washington, U.S.A.

The Economic Development Association of Skagit County (EDASC), Washington, is focusing on businesses that "fit community needs," according to Kathleen O'Brien of O'Brien & Company, a recycling and waste management research firm based in Bainbridge Island, Washington. Kevin Morse of EDASC provided the following comments in an article by O'Brien in *In Business* magazine (November/December 1996):

> Each community that is interested in setting up an eco-industrial park has to set up its own self-designed criteria that serve the community needs. There has to be a community fit that brings companies that can relate to "unsolved" community needs. In our case, a particular company can provide a solution to our biosolids disposal problem. We have many shellfish industries here on the coast in northwest Washington and they need uncontaminated water. We also have large farms raising livestock, and they have manure disposal problems. The company that we are working with now to locate in our eco-industrial park can turn waste products into animal feed as well as fertilizer.

O'Brien & Company, Hovee & Company, and Shapiro & Associates comprise the consulting team for the Skagit County eco-industrial park. Here is their vision for the project as defined by a recent EDASC-sponsored planning and feasibility study:

> a place where recyclers together with firms and organizations who pro-
> duce, sell, distribute, serve or educate with goods and services that are
> environmentally friendly can work together for profit, jobs, and envi-
> ronmental enhancement.

According to O'Brien, the Skagit County project will include a Resource Recovery Center, a Manufacturing Center, a Community Center, a Sales/Marketing Center, and an Environmental Business Center. The Resource Recovery Center, providing a central location for cost-effective recovery of materials, will collect, recycle and process residual flows and by-products from the community and participating businesses and act as an equipment "bank" for sharing forklifts, trucks, scales, and other materials handling "tools of the trade." The Manufacturing Center will include existing and start-up "green" companies, plus more traditional companies that commit to a "zero emissions" goal. The Community Center will offer support demonstrations of green technologies and organic farming techniques and coordinate community-wide events that support the eco-industrial park. The Environmental Business Development Center will provide a number of technical services including: technology and feedstock brokering; research and development; business support equipment and supplies; and space for environmental and planning professionals. These centers collectively form an enterprise community that supports the needs identified by Skagit County citizens.

These examples also show a variety of strategies being applied, and tested, to develop business ecosystems. Some may succeed. Some may fail. This is to be expected. Nature often experiments with different approaches until a successful one is found.

These and other business ecosystems are evolving organizational systems that reflect different needs, resources, and approaches. In the context of business ecology, their development is affected by both their social DNA—core purpose and values—and their business environment. **The more closely aligned the core purpose and values, the better the chances for a good fit among participating organizations.** Each business ecosystem, for example, is shaped by the people and organizations involved, the type of economy, proximity to natural and human resources, the regulatory climate, and organizational capability.

LEARNING FROM AGRO-ECOSYSTEMS

While closed-loop behavior is novel and exciting in the manufacturing sector, it is somewhat "old hat" in the agricultural and food sector, particularly for those of us who are familiar with "organic" or sustainable agriculture. For that matter, anyone who has raised their own food and managed a compost pile appreciates life-sustaining flows such as water and nutrients, how the garden is rejuvenated the next year with compost developed from previous years' residual crops and food by-products, and the value of co-planting. For instance, oregano, tomatoes, and marigolds together thwart pests and encourage healthy growth. A community garden, an agro-ecosystem on a small scale, can be a powerful model for understanding business ecosystem development.

In fact, the industrial sector has a lot to learn from its older predecessor, agriculture, in particular, the ancient and highly evolved agro-ecosystems that preceded "modern" industrial agriculture. The following examples provide ample proof that there is much to gain by merging these ancient techniques with today's technology. By rediscovering these techniques, we can develop healthy, life-sustaining companies within sustainable, ecological economies.

Agro-Ecosystems in Asia, the Middle East, and South America

Joseph Levine and Kenneth Miller describe in their textbook *Biology: Discovering Life*, how Western agriculturalists are rediscovering agro-ecosystems from Egypt and Asia:

> But as Western aquaculture technology has grown, researchers have rediscovered many remarkably efficient culture techniques known in Egypt and Asia for millennia. A single agriculture-aquaculture farm in Thailand, for example, can support several acres of crops; thousands of ducks, chickens, and pigs; and a pond with over a million fishes [Figure 5.2]. The secret is intensive internal recycling; crops are fed to animals whose wastes fertilize pond water to grow algae and other fish food. Water from fish ponds is used to water and fertilize crops simultaneously, and other wastes produced by the system are either recycled into fertilizer or digested by bacteria to produce methane gas for use as fuel.

The dynamic system of exchanges among the "organisms" shown in Figure 5.2 is a "business ecosystem" that is supported sustainably by solar

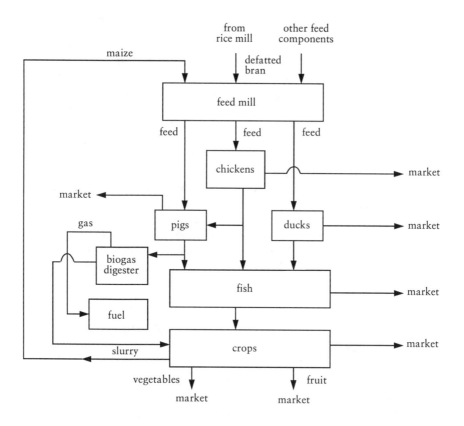

Figure 5.2   Diagram of an Agro-Ecosystem in Thailand
Source: *Biology: Discovering Life*

energy and other natural resources. Crops use sunlight, water, nutrients, and carbon dioxide to "photosynthesize," grow, and form the base of the "food web." Animals, in turn, ingest the plants and breathe oxygen to support their metabolism, and expel waste to be recycled back into the system or converted into energy.

In our industrial economy a food web also exists. However, its base is primarily fossil fuels, the product of many millions of years of previous biological activity, which was also driven by solar energy. Plastics, gasoline, pesticides, and other products use petroleum and other "fossil" resources to drive their industrial metabolism. Levine and Miller point to the obvious dilemma posed by a growing, energy-intensive industrialized economy and provide some practical advice:

Developed countries consume enormous quantities of energy today, and as Third World countries industrialize, their energy demands will rapidly outstrip ours. This trend is important because most forms of energy have become significantly more expensive in the last half-century and will continue to increase in price. The reasons for this are simple. In the early days of the Industrial Revolution, low energy prices reflected only the low costs of production. Now we must factor in not only increased costs of finding, extracting and transporting energy supplies but also the expense of controlling pollution.

Today more than 80 percent of the world's energy comes from non-renewable energy sources, which include finite deposits of oil, coal, and natural gas. The fact that these reserves are finite and dwindling cannot be emphasized strongly enough. No matter how much remains today, and no matter how efficiently we extract and conserve it, our supply will run out sooner or later. Analysts warn, for example, that little of the world's petroleum will be left by the middle of the next century. The transition from fossil fuels to other energy sources will be much smoother if we prepare for it before current supplies run low.

Here is another agro-ecosystem example provided by Dr. Vu Quyet Thang of the National University in Hanoi, Vietnam:

Farmers in Vietnam are increasing the productivity and sustainability of their back-yard farms with help from the Vietnamese Gardeners' Association. The association is promoting VAC—a process of mixed cropping that provides crops with improved nutrition, nourishes the soil, and provides cash crops. The VAC program is partly sponsored by UNICEF and promotes integrated farming of vegetables, pigs and fish.

A farmer in Xuan Phuong, outside Hanoi, is growing vegetables in his 720-square-metre front yard for direct marketing in the several open markets of Hanoi. The yard grows grapefruit, oranges, bananas, papayas, sapodila, mint, squash, onions, amaranth, protein-rich sauropus and sweet potatoes. The plants grow at different levels and heights, providing shelter, shade, and nutrition to each other. The leaves of some plants are fed to pigs, while other parts—such as the roots of sweet potatoes—are for human consumption. The yard is fertilized with pig manure and human waste.

The farm includes a small fishpond that has about 1,500 fish. Species are carefully chosen to be symbiotic. Tench tend to feed near the top of the pond, carp in the middle and tilapia at the bottom—feeding on the waste of the fish species living above it. The pond is covered with water hyacinth, which provide oxygen for the fish, protect them from the sun and are used to feed the pigs. The yard also has a pigsty with a sow that produces up to 20 piglets a year. Yard greens and fish-laced meal are served to the sow.

In 1992, one farmer using the VAC system made about US$450 from his yard. (The average Vietnamese annual income is US$240.) UNICEF estimates the income of VAC farmers to be from three to ten times higher than rice farmers.

VAC gardens are also established by schools, churches, orphanages, old-age centers, and factories in Vietnam, providing their users free or subsidized nutritious food. The gardeners association has a corps of extension workers who are experts in the various technologies and farming systems and who provide extensive and regular advice to the VAC farmers.

This example conveys a number of lessons for business ecosystem and sustainable development. First, we have a lot to learn from other cultures who, by some Western standards, are "backward" or "not modern." **Respect for other cultures, and the ability to see from different perspectives, is critical to sustainable development.** Where possible, we should use proven techniques, and, as necessary, adapt these techniques to local conditions. New approaches, when needed, should integrate both proven, cultural practices and "modern" knowledge and technologies. T. V. R. Pillay, a recognized aquaculture expert from India, discusses "appropriate technologies" in Chapter 2.

The Vietnam example also illustrates how diversity builds resiliency, system "robustness," and productivity. It also shows the interdependence of "organisms" and the benefits of many closed-loop systems that are fully integrated into the natural ecology. Further, in the context of global sustainability, "food" should be a natural currency available to all citizens, regardless of social or economic class. And, in this light, we are interdependent "organisms" sharing a global ecosystem. It is in everyone's best interest to find ways of balancing all human needs with those of other creatures that share our biosphere.

Ignachy Sachs and Dana Silk discuss the role of integrated food-energy systems (IFES) in *Food and Energy: Strategies for Sustainable Development* (1990). Here they describe the work of The Food-Energy Nexus Programme (FEN) of the United Nations University:

> One of the two major focuses of FEN was the analysis of systems designed to integrate, intensify, and thus increase the production of food and energy by transforming the by-products of one system into feedstocks for the other. Such integrated food-energy systems (IFS) can operate at various scales, ranging from the industrial-sized operations of

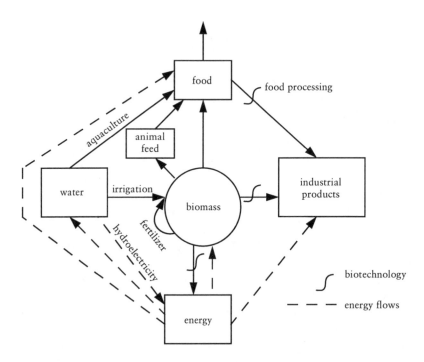

Figure 5.3   A Diagram of Integrated Food-Energy Systems
Source: *Food and Energy: Strategies for Sustainable Development*

Brazil designed to produce primarily ethanol and fertilizer (La Rovere and Tolmasquim 1986), to the village and even household-level biogas systems of India (Moulik 1985). A conceptual outline of such systems, including the long-ignored but vitally important component of water for biomass-based production, is shown in Figure 5.3.

This valorization of agricultural by-products plays an important role in the design of IFES for specific agro-climatic regions, designs which should closely follow the paradigm of natural ecosystems. These modern, ecologically sound systems are characterized by their closed loops of resource flows. In a sense, they incorporate the same rationality of traditional farmers, albeit at a completely different level of scientific and technological knowledge.

The study of such systems, adapted to the diversity of natural environments and responding to a wide spectrum of needs in terms of size, technological sophistication, and capital intensity, became the main thrust of FEN research activities in this field. Earlier UNU work on patterns of resource use and management in Asian villages (Ruddle and Manshard 1981) provided a very useful starting point for this research.

Sachs and Silk provide a specific IFES example, from the community of Camacan, in Bahia, Brazil, that integrates a petrochemical complex with regional agriculture and illustrates the need for systemic planning:

> The 1984 Brasalia seminar described IFES as the adoption of agricultural and industrial technologies that allow maximum utilization of by-products, diversification of raw materials, production on a small-scale, recycling and economic utilization of residues, and harmonization of energy and food production. Such systems imply the need for comprehensive land-use planning as well as planned interrelations among soil, water, and forest resources in relation to agricultural residues. Major advantages are their minimal negative environmental impact, and their decentralization and efficiency, which often have positive social and economic side-effects.
>
> In Brazil, IFES notably include micro-distilleries for alcohol production from sugar cane, sweet sorghum, manioc or sugar beets; fattening of stall-fed livestock biodigestors for decomposition of livestock manure and/or sugar cane bagasse and/or stillage; generation of electricity and agricultural mechanization on the basis of fuels thus produced; and the application of biofertilizer to cultivated lands. La Rovere and Tolmasquim (1986) note the possibility of extending this to include various ways of enhancing the water management system, including the production of aquatic plants, fish, and zooplankton as well as the retention of water for sedimentation and irrigation purposes.

Bioshelters, Inc., Amherst, Massachusetts, U.S.A.

Bioshelters, Inc. in Amherst, Massachusetts, consists of several closed-loop businesses in one. Founder John Reid combines aquaculture, hydroponics, and solar engineering into a profitable business ecosystem that produces 600 pounds of fish and 50 to 60 cases of basil each week. This translates into about 30,000 pounds of fresh market-proven fish and 4,680 cases of basil per year. The recirculating, self-contained modular system has several attractive features. The hydroponic system is directly linked to the aquaculture systems, making fish wastes available as fertilizer while cleaning the aquaculture tanks. About 99.7 percent of all its water and waste is recycled, making the business extremely lucrative in an area that has water shortages and stringent effluent regulations. It can also be located in almost any climate in close proximity to its target market. In the context of business ecology, it has a broad-ranging habitat, can position itself favorably with customers, and gains advantage over competitors by avoiding competitors' shipping and preservation costs.

Bioshelter, Inc. raises tilapia, a fish originally from Africa, for several reasons. First, tilapia grow quickly, reaching one-pound weight within six months. Second, they eat little food to reach this one-pound market size. According to Reid: "It takes one and two-tenths pounds of feed to make one pound of fish over the fish's six-month life." Reid contrasts this with cattle, which require about 30 pounds of feed to produce one pound of beef. Third, tilapia is very marketable in the U.S. with a 30 percent annual increase in consumption.

Racks of basil plants hang over the tilapia tanks. The basil grows on the water, extracting nutrients from the fish excrement while helping to filter the water. Tomatoes, broccoli, and other produce are also grown in the greenhouse shelters. The fish and plants are raised without added fertilizers, pesticides, or fish antibiotics. In addition, Reid has created synergies with the local community by employing physically challenged citizens and serving as a model for ecologically-based agriculture.

DEVELOPING HEALTHY BUSINESS ECOSYSTEMS: CHALLENGES AND OPPORTUNITIES

The previous sections highlighted examples of business ecosystems from around the world and illustrated how much we can learn from agro-ecosystems. As discussed in Chapter 3, shopping malls and other forms of development are also business ecosystems. However, while business ecosystems are ubiquitous in our society, most are not fully formed, sustainable, or genuinely closed-looped. What are some of the challenges and opportunities related to closed-loop economic development? How can we develop healthy, sustainable business ecosystems?

A Strategic Vision for Development

CHALLENGE—*Seeing and Removing Obstacles*   Creating healthy business ecosystems requires a new way of seeing, such as the ability to see systemic relationships, patterns, and flows from different perspectives. Seeing in this way can spotlight both obstacles and openings for developing vital flows and relationships to support sustainable development. In fact, while business ecosystem development offers significant opportunities, there are potential obstacles that exist at various scales. These include:

- Do local zoning codes prohibit the co-location of businesses that might share resource flows?
- Are Federal laws, such as the Resource Conservation and Recovery Act, so restrictive and complicated that they will reduce pollution prevention, technology innovation, and recycling opportunities?
- Will lending institutions support innovative technologies and recognize the synergistic benefits of business clustering, such as reduced operating costs, collaborative production, marketing, and distribution?
- Will a community accept the changes or novel approaches?
- Do participating organizations and individuals have ample skills and training?

OPPORTUNITY—A Systemic, Strategic Planning Tool   Business ecology is both a lens and an organizing framework for developing business ecosystems. The ability to see from different perspectives, as discussed elsewhere in this book, is crucial to understanding and developing sustainable human systems, including organizations, technologies, and economic systems.

The Business Ecology Network (see Appendix) has developed the Business Ecology Roundtables^{SM} process, a series of workshops, mindware—organizational models based on natural systems—and tools, to help business and community leaders envision and create healthy business ecosystems. The process starts by identifying values, visions, needs, resources, and core purpose, and strategically assessing the business environment, or the context for development. **Potential development is viewed through five "system" lenses—1) society, 2) science and technology, 3) economics and financing, 4) policies and institutions, and 5) the environment—to develop a holistic picture of the evolving business environment, including potential obstacles and opportunities.** A strategic development framework emerges, which includes how to build organizational capacity, develop vital flows and relationships, create opportunities for sustainable enterprise, and manage resource exchange networks.

Learning and applying business ecology through the Roundtables process creates a systemic, viable development plan at all scales of development, including company, community, and regional business ecosystems.

Merging Ecology and Commerce

CHALLENGE—Seeing Beyond Cash Flow   Many businesses see profit as an essential means to achieve broader purposes and more enduring success. Others see profit as *the* purpose of business. Whatever the case, most decision-makers find that their actions are driven by the bottom line, with maximizing profits for shareholders as an overriding objective. This bottom-line focus has at least three consequences:

- The "cash is king" mentality obscures or downplays the importance of other value-creating flows such as information and ideas, people and other organisms, air, food, energy, water, materials, and products and services.
- Other stakeholders, such as customers, suppliers, communities, and the environment, often take a "back seat" to the shareholder or management, even though they significantly affect, and are affected by, a company's decisions and actions.
- Decisions driven by the bottom line may often conflict with the values of a decision-maker or an organization.

Development of a sustainable business ecosystem requires that participating organizations have a more balanced, systemic perspective of themselves and their environment. Like organisms, organizations must be keenly aware of all vital flows and relationships that sustain them. Participating organizations, or an organizing agent acting on their behalf, require this ecological knowledge of individual firms and an understanding of the local and regional ecological economy to create closed-loop resource flows that support a robust business ecosystem.

OPPORTUNITY—Using Systems Thinking to Develop Vital Flows and Relationships   While cash flow and profitability are essential to your company, the lens of business ecology presents them as part of a broader system of life-sustaining flows and conditions. Viewing your company as an organism can improve efficiency and prevent pollution in the short term and thus save money and improve your bottom line. **A comprehensive application of business ecology redefines organizational viability to include all life-sustaining flows and stakeholder relations. This is relevant to individual businesses and organizations as systems of companies, a business ecosystem.** A bottom-line-driven organization often misses opportunities for improving ecological efficiency and viability by focusing exclusively on cash flow. Business ecology, by expanding your organization's perspective from "cash flow" to "life flows" and "shareholders" to "stakeholders," can help you discover these other dimensions of efficiency and viability. Business ecology values all wealth-creating flows sustaining an organization, such as: information and ideas; money, people and other organisms; air, food, energy, water, products and services, and materials. All these flows support the metabolism of a company within its environment. Business ecology also values stakeholder relations; it recognizes the web of people and organizations that support and dynamically interact within an organization. This web includes: customers, shareholders, creditors, suppliers, manufacturers, distributors, utilities, retailers, citizen and community groups, various levels of government, contractors, accountants, lawyers, and competitors.

Business ecology's organic model can help you assess and develop the vital flows and relationships that sustain your organization. This, in turn, helps you improve the viability and ecological efficiency of your organization while determining how it "fits" within developing business ecosystems. Business ecology provides a systemic lens to help you discover your organization's metabolism, niche, and habitat within its broader ecological economy.

Values-Based Organization and Development

CHALLENGE—Building Organizational Capacity As the examples in this chapter show, a strong social fabric, shared values, demand for services/resources being exchanged are critical to creating successful business ecosystems. In the case of Kalundborg, the business leaders socially interact on a regular basis, building trust, openness, and cooperation. In such a climate, genuine collaboration is rewarded and ultimately profitable for all involved. The success of future business ecosystems will hinge largely on building organizational capacity and finding the right incentives to encourage inter-firm resource exchange and collaboration.

OPPORTUNITY—Aligning Development With Core Values and Purpose
Successful business ecosystems, such as Kalundborg, Denmark, are shaped by values that guide decisions and actions, build community, and foster communication—all essential to a viable system of organizations that exchange energy, materials, water, and other flows. Identifying and aligning core purpose and values—or social DNA—is, in fact, critical to healthy business ecosystem development. This social DNA defines a business ecosystem's identity, shapes the development of vital flows and relationships, and strengthens its organizing capacity.

Business ecology, as a values-based organizing framework, helps business ecosystems become self-organizing. Shared values and purpose—social DNA—articulated by customers, communities, businesses, and other stakeholders acts as a genetic code for development. Because it is based on natural systems design, business ecology is fractal; it works at different scales of organization, such as a small business, community, a nonprofit organization, and regional, national, and international enterprises. In essence, **business ecology strengthens the self-organizing capacity of developing business ecosystems.**

A Flexible Organizational Model

CHALLENGE—Accommodating Diverse Organizations and Needs   Organizations that wish to participate in business ecosystems often vary in size, type, and level of understanding. Further, different communities and regions may have very different needs, resources, and capabilities. A successful business ecosystem must accommodate this diversity and build on shared strengths.

OPPORTUNITY—An Adaptable, Easy-to-Use Framework   Business ecology's organic model for organizational management and design and Roundtables^{SM} systems thinking can assist businesses and organizations with varied requirements to develop and/or participate in business ecosystems. Business ecology helps organizations, regardless of scale or complexity, identify the common purpose, shared values, stakeholders, and resource exchange networks essential to business ecosystems.

An Organizing Framework for Sustainable Development

CHALLENGE—Realizing the Opportunities of Sustainable Development
While there are many opportunities for integrating economic, social, and environmental goals in the current business environment, more opportunities can be realized by developing a sustainable, ecological economy. Our industrial economy of mass production, marketing, and consumption creates several conditions that run counter to healthy business ecosystems, communities, and lifestyles:

- Overconsumption, that is, a lifestyle where people acquire material wealth to the point where their quality of life actually diminishes;

- Dependency creation, overspecialization, and consumerism that diminish our human potential, including creativity, self-reliance, ingenuity, and sense of community;
- Natural resources, people, and other factors of production, or wealth creation, are perceived as inexhaustible and replaceable and often undervalued through monetization; and
- Separation of the production of products and services from where they are consumed or used to the extent that closed-loop thinking is a lost art; this includes everything from life-cycle product design to simply knowing where things come from and where they go.

Economic policies, whether they are publicly or privately driven, can create a healthier environment for sustainable enterprise by introducing incentives for closed-loop behavior and improving ecological efficiencies. In such an environment, profitability, stakeholder relations, and environmental performance work together—creating win-win-win outcomes for business, society, and the environment. In fact, as examples in this chapter show, financial, tax, and legislative incentives often stimulate business ecosystem development.

However, while these incentives make sense to many stakeholders, they may not be readily accepted by those who see these changes as wealth transfers away from them. Learning from the lessons of pollution prevention, full-scale implementation of tax and subsidy changes must involve all stakeholders, including, for instance, accountants, tax lawyers, marketing professionals, and others who perceive that their livelihood may be threatened by these systemic changes. Fortunately, there are growing numbers in these professions who see the challenges and opportunities of sustainable development. To the extent possible, sustainable solutions must be seen as a "win" for everyone, and be based on incentives and timetables that allow as smooth a transition as possible.

OPPORTUNITY—A Powerful Catalyst for Sustainable Development

Business ecology is a powerful catalyst for sustainable development. Its models for organizational management and design are mindware for sustainable enterprise; these models help organizations create value that genuinely improves the quality of life and optimizes value creation for all stakeholders; they help organizations adapt and form collaborative networks while reconnecting the economy, communities, and the environment. By using an ecological lens, leaders, businesses, and organizations can create closed-loop solutions, where residual flows and "waste" become new opportunities for sustainable enterprise. This lens can help you see essential, life-supporting services, unmet needs, and resources that are often not captured or undervalued by bottom-line thinking and the industrial economy (see Chapter 2). For instance, it takes into account services provided by natural systems, such as flood mitigation by wetlands or filtering of water by oysters, and vital human services, including building healthy communities, nurturing children and families, and providing meaningful, life-long education. How can we recognize both quantitative and qualitative indicators of success? The need to balance and link these measures? Business ecology recognizes that developing human potential and successful organizations means supporting the development of enduring, meaningful work, vibrant communities, and healthy business ecosystems. The next chapter describes how building community and strong stakeholder relationships can help your organization thrive within its business ecosystem.

6

The Ties That Build Community— and Profits

In and through community lies the salvation of the world . . .
M. Scott Peck

Community is an expression of nature's organizing elegance. Community is essential to our lives and who we are. It reinforces a "sense of belonging," a deep sharing of values, vision, purpose, and commitment. It is also about "place," and our myriad personal, professional, and cultural connections with individual places. A shared sense of community lies at the heart of a healthy, life-sustaining organization and contributes to our spiritual well-being.

Business ecology, as a framework for sustainable enterprise, puts "community" back in business. What is a business's community? It is its connections and relationships—its stakeholders. This chapter highlights leaders, businesses, and organizations that are building strong stakeholder communities and becoming more viable, profitable, and competitive. Companies, such as Ben & Jerry's, Whole Foods Market, Shaman pharmaceutical, and Arm & Hammer are reaching higher levels—including sustained earning streams—by building healthy relations with their employees, customers, suppliers, and other stakeholders. This chapter also includes profiles of successful community-based organizations, such as the World Bank Spiritual Unfoldment Society, and Food from the Hood in Los Angeles, California, which are choosing, and developing, sustainable paths to the future.

ALIGNING VISION, VALUES, AND PURPOSE

Everywhere there is a hunger for a new paradigm, a hunger to be connected, to be a part of something bigger, something that challenges each of us to be truly alive and have purpose. Most of us sense that the rigid, mechanistic design, industrial-age organizations are life-depleting, obsolete, and ill-suited for today's rapidly evolving business environment. **We want organizations that recognize not just what we do, but** *who we are.* We want organizations that are living organisms that nurture our creativity, ingenuity, and sense of purpose. We are longing for community, where our lives and work are fully integrated with our vision, values, and purpose.

Margaret Lulic, author of *Who We Could Be at Work,* provides stirring accounts of such struggles and triumphs within the business community. In an example from her book, Sue Anderson left corporate life to become executive director of the Bloomington Arts Center. Now, in a challenging and creative position, she looks back at her previous experiences:

> I think companies take away part of a person's spirit, their lust for the job. We get stifled, we shut down, and then we don't work as well. I started shutting down my creativity as well as my output. I became dissatisfied with myself and with what I was accomplishing.

As Lulic points out, Anderson's former employers lost more than an executive secretary or administrator. They lost a person with considerable energy, skills, and knowledge, who was unable to develop fully her creative potential.

Unfortunately, not everyone has taken the initiative to move from an unfulfilling environment to discover the kind of purpose, creativity, and fulfillment that Anderson currently enjoys. Here's another profile from *Who We Could Be at Work,* a vice president of finance and human resources for a subsidiary of a Fortune 500 company. She tells her story anonymously while still inside her organization:

> I've experienced many types of organizational change in my company over the past 15 years: the entrepreneurial growth phase, the changing of the guard from a more technical/business entrepreneur to a business/marketing-oriented person, an acquisition by a much larger organization that now has been acquired by an even larger industrial organization. My experience with these changes has taken a heavy toll on me and the other employees of this company. I don't think it was the number of changes but the sense that we were going one step for-

ward and two steps back, especially with the last few changes. It's hard to keep up your commitment when you suspect all your efforts will be reversed.

The most important pattern I've noticed through these experiences is how much difference it makes when the employees feel they have control. When we felt we still were in the driver's seat, we remained focused and personally committed. Every decision that took away control ate away at people, and what we really lost was each person's sense of personal responsibility and enthusiasm.

These examples illustrate what Lulic describes as "the deplorable waste in our current organizational paradigms." What can we do to bring life back into our organizations? Help people feel they are valued contributors? Keep employees focused, committed, and motivated? Certainly a critical step is recognizing that **the current systems and designs are not working because they do not align vision, values, and purpose.** Articulating a personal or organizational vision is a powerful motivator. It gives an individual or organization a common reference, a place or outcome to be reached. Work that is purposeful and meaningful brings the best out in people. In fact, when vision, values, and purpose are aligned, people and organizations can achieve amazing results. Work becomes an energizing, joyful experience, an expression of our creative potential, our true life's calling and deeper purpose.

In *Find Your Calling, Love Your Life—Paths to Your Truest Self in Life and Work*, authors Martha Finney and Deborah Dasch follow the lives of over a dozen men and women who have discovered their callings. Their jobs range from airport director, animal cruelty investigator, and hairstylist, to organic herb gardener and mystery writer, but there is a thread that brings them together: Finney and Dasch's conviction that "Our true life's work returns us to ourselves." The authors mark guides along the route to finding one's calling, such as "trust your basic nature," "stay flexible and open," "turn handicaps into assets," "reject fear," and "engage both your head and heart."

James Hardeman, Head of the Employee Assistance Program for Polaroid Corporation, says this in *Find Your Calling*:

Once you realize there are others who share your vision and are just as committed as you are to turning that dream into a reality, the right doors will open, one by one. We know intellectually that to choose job security over our values is a bad trade. We all know that job security is a rapidly disappearing illusion. But would we have the strength to make the right choice when we must weigh our morals against our mortgage?

Myra Doms, an outplacement counselor interviewed in *Find Your Calling*, sums up what many are already discovering for themselves: "The whole nature of work is being changed. It's not necessarily negative, you just have to find your new place."

Clearly there is a hunger for a new paradigm, for spiritual meaning, connection to self and community, and purposeful work. In the ecological economy, building meaningful connections and relationships is essential. Within organizations, individuals need an environment that cultivates self-knowledge and the discovery of one's purpose and values. Then an individual's and an organization's place in the web of life will unfold naturally and successfully.

Figure 6.1 is a self-referential map of organizational values. This systemic tool can help you track progress toward identified values and evaluate or anticipate the impact of decisions. For the purpose of illustration, the seven core values of the Business Ecology Network (see Appendix) are plotted. Together with our core purpose, "To be a catalyst for life-sustaining enterprise," these values describe our learning community—who we are and what we do. Each of the plots on the radii measures how well our organization is aligned to each of our core values, our vision, and our purpose. This is an effective filter for identifying and assessing:

- Organizational goals
- Product design
- Project development
- Short- and long-term strategies

For instance, does a new project reflect our core values? The seven value "spokes" are spaced along equal arcs of the circle to convey an ideal sense of balance. A combination of quantitative and qualitative indicators and techniques are used to evaluate the Business Ecology Network's progress toward each value. These include, for instance, questionnaires among employees, surveys of customers and suppliers, and detailed assessment of flows and processes that create value within the organization. Collectively, the polygon is a composite picture of how well our organization is aligned with its core values. This "values polygon" effectively correlates values and goals for individuals, businesses, organizations, and larger systems, such as business ecosystems.

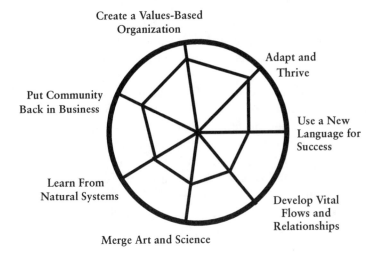

Figure 6.1   Is Your Organization Aligned With Its Values?

© 1998 Business Ecology Associates

A NEW MEASURE OF ORGANIZATIONAL SUCCESS: ACCOUNTABILITY TO STAKEHOLDERS

A company's purpose, responsibility, and accountability extends well beyond profit for the shareholders. A company clearly impacts, or is impacted by, a broader community of stakeholders in ways that have both quantitative and qualitative effects.

Stakeholders include all those people and organizations that affect, and are affected by, a company's existence. These can include customers, shareholders, creditors, suppliers, manufacturers, distributors, utilities, retailers, citizen and community groups, various levels of government, contractors, accountants, lawyers, and competitors. For example, actions such as downsizing, outsourcing, or moving an industry offshore for cheaper wages or less environmental restrictions may provide a short-term benefit for some but not all stakeholders. There may be long-range economic, social, and environmental effects, such as underemployment in former industrial areas, lack of employee health care coverage, and abandoned factories, even hazardous waste sites. Many business leaders already have a broader sense of purpose and responsibility, but they are constrained by an inflexible bottom-line accounting system that rarely allows them the latitude to make decisions that reflect their deeper values. Ralph Estes, who served as chief financial officer of several corporations, calls this: "the tyranny of the bottom line."

Estes worked as a certified public accountant auditor with Arthur Anderson & Co. and was a past president of Accountants for Social Responsibility. He is also a professor of business administration at The American University, as well as resident scholar and cofounder of the Center for the Advancement of Public Policy in Washington, D.C. The Center, as a nonprofit research and educational organization, works to create a more rewarding, humane work environment and greater corporate accountability to stakeholders. Here is an excerpt from Estes' book, *Tyranny of the Bottom Line*, that conveys the need for a new, stakeholder-based scoring system for business:

> Business can be run successfully, with a fair return not only to shareholders but to all stakeholders, and still be humane. It can be competitive, it can be financially strong. And it can be fun.
>
> To make it that way, we will first have to understand what really drives the corporation—what it is about business that makes good people do bad things. We will see that it is an archaic scorekeeping system that does not do what it is supposed to do: it doesn't measure the real performance of the firm.
>
> Once a problem is accurately diagnosed, the prescription often becomes obvious. We need a fair scorekeeping system, one that will send different messages to corporate managers. A scorekeeping system that simply shows the effects of a corporation's actions on all its stakeholders, not merely its stockholders, and then tells managers that they will be responsible for these effects.
>
> *Tyranny of the Bottom Line* lays out a prescription in an effective and workable program that can make corporations safer and more rewarding for all of us, and more enjoyable, more honorable, for the people who run them.

Meeting the needs of all stakeholders can be best accomplished by understanding and working with the goals that currently motivate business leaders. As Estes points out, under the present system, corporate managers are motivated to maximize the bottom line for shareholders, often at the expense of nonshareholder stakeholders who have vested interests in the company. Developing sustainable enterprise requires a new measuring system for organizational success that explicitly recognizes the viability or health of all stakeholders.

Why are stakeholder relations important? Companies that understand and develop healthy stakeholder relations will be the leaders of the knowledge-based, ecological economy for two important reasons:

(1) successful companies, such as Interface, Inc. and the Monsanto Company, already see the value of sustainability and healthy stakeholder relations and are building strategic alliances to gain competitive advantage; and

(2) most stakeholder communities are exerting greater and greater influence. This trend will likely increase in a knowledge-based economy.

Leading companies, such as Malden Mills (see Chapter 4), demonstrate convincingly that profitability, social responsibility, and environmental performance can work together, particularly when a company cultivates healthy stakeholder relations. Many companies are already recognizing their broader social purpose and responsibilities in this regard. One is Ben & Jerry's, the Vermont-based ice cream company. They have been working closely with Ralph Estes and others to develop a broader, more meaningful accounting system. Here is the introduction from Ben & Jerry's *1994 Social Performance Report* written by author and business leader Paul Hawken:

> How does society benefit from the activities of business? More to the point of this report, how can we measure the "profitability" the community gains from a company like Ben & Jerry's? From its outset, Ben & Jerry's has pioneered in its unwillingness to accept the premise that a growing, profitable business is in itself a social good. For seven years, it has asked outside consultants to conduct a social performance assessment, an honest and candid look at the company's impact on workers, the community, and the environment.
>
> Measurement of social impact is not easily objectified. The assessor makes value judgments without benefit of any "generally accepted accounting principles." In the end however, the arbiter of a company's social impact is people. What do employees think? What do customers experience? How do vendors judge their interaction with the company? What is the local community's relationship with the company? Such questions cannot be measured precisely. Although somewhat calculable, what works best is to measure changes from year to year. Is the workplace more or less safe? Is morale up or down?
>
> In that light, 1994 was a year in which Ben & Jerry's showed dramatic improvement is some areas, good work in others, and slippage on some key issues. Improvements in safety, quality, environmental

awareness, employee equity, and community involvement are substantive and are, in some cases, a reflection of how the Company has responded forthrightly to past social assessments. Where the Company has not improved its social performance, in areas such as employee morale, there are deeper and longer-term issues that have come to the fore in the past year. Improvement in one area can illuminate the need for improvement in other areas. Yet even in areas where Ben and Jerry's needs improvement, the standards being applied are unusually high. For example, where employee morale at Ben & Jerry's declined in 1994, morale overall is significantly higher than the corporate norm. In other words, what is a problem at Ben & Jerry's would, in many other places, not even show up on the radar screen. The Company has set a tough task for itself: lofty goals, high standards, and great expectations. . .

Another company, Whole Foods Market, the largest natural food supermarket chain in the U.S., is committed to enriching the lives of all its stakeholders: customers, team members, shareholders, and the larger community. The company measures its success by how well it meets the needs of these stakeholders. This includes, for example, building strong relationships with its customers, creating a system of local supply "webs" which include organic farmers, community groups, and natural healthcare suppliers. In its 1996 *Annual Stakeholders Report,* Whole Foods Market describes its vision and core values:

> Sixteen years ago, the founders of Whole Foods Market combined their love of good food and vision of right livelihood to create a one-stop shopping experience for natural foods in Austin, Texas. Since that humble beginning, Whole Foods Market has grown into the nation's largest chain of natural food supermarkets by offering our customers the very best natural, organic, gourmet, and ethnic products available. Our wide array of high quality foods and our customer service provide a unique and gratifying shopping experience.

Whole Foods Market's commitment to its vision, values, and stakeholders has paid off. Beginning with its first store in 1980, the company grew to 68 stores in seven regions and sales of $892 million in 1996. The company plans to have 100 stores across the U.S. by the turn of the century.

Investors, employees, communities, religious groups, and other stakeholders are already leveraging financial resources to ensure greater corporate accountability and social responsibility. In a sustainable, knowledge-based economy this will likely increase as people have greater access to information through the Internet and other information technologies. Better informed, well-organized stakeholder communities are already working to help companies become more aware of the impacts their actions have. For example, the Jessie Smith Noyes Foundation in New York City leveraged its shareholdings and worked with the Southwest Organizing Project to help community groups participate actively in the siting of an Intel manufacturing facility in Rio Rancho, New Mexico. Prior to the Noyes Foundation's involvement, Intel was proceeding without community involvement. Companies, large and small, are realizing that it is in their best interest to be good corporate citizens. In fact, many businesses see that healthy stakeholder relations, as the examples in this chapter illustrate, create more viable, profitable organizations and stronger communities.

How can you measure the health or viability of your organization's stakeholder relations? Business ecology recognizes that both quantitative and qualitative indicators are important. Figure 6.2 is a systemic tool for assessing your company's multiple stakeholder relations simultaneously. This example plots shareholders, employees, customers, suppliers, communities, the general public, and the environment on a circle to represent the stakeholder community. These seven plots are equally spaced to convey an ideal sense of balance among the several stakeholders. The points on each of the radii inside the circle—"stakeholder spokes"—are an overall measure of the health of each stakeholder relationship. Collectively, the shape or polygon shows how a company is doing with respect to the ties that build its stakeholder community. That is, points closest to the circumference are the strongest, while the points closer to the center indicate areas that are weak and underdeveloped.

The points on each of the "stakeholder spokes" can be determined from a variety of quantitative and qualitative factors. For example, shareholder health can be measured by traditional means, such as net income, earnings per share, return on investment in assets, and return on shareholder investment. It can also be measured by qualitative data such as surveys that ask investors to rate a company's performance with respect to the investor's—or society's—economic, social, and environmental goals. For example, does the company maintain adequate safety, quality assurance, or environmental performance programs to protect the investor from unreasonable risks or liabilities? All of these factors can be combined into a cumulative score that expresses a company's overall performance to share-

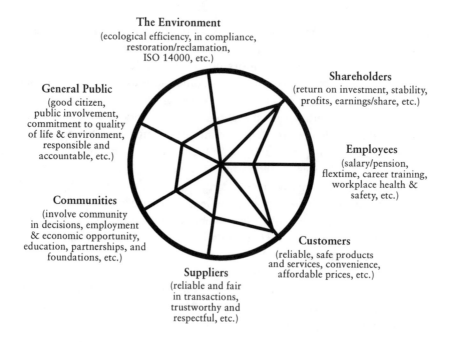

Figure 6.2   How Strong Are Your Organization's Stakeholder Relations?

© 1998 Business Ecology Associates

holders, for example, as a percentage or on a scale of one through ten. Each factor could also be expressed in terms of meeting investors' goals.

Employees, on the other hand, may consider the following as important: salary and advancement opportunities, profit and equity sharing programs, workplace health and safety, the company's employment record, employee grievances, impacts of technologies developed by the company, pension programs, childcare, flextime, and employee involvement in company planning and decisions. Like the shareholder health indicator, the employee health indicator plotted on Figure 6.2 is a composite of several of these quantitative and qualitative measures. Similarly, other stakeholder health indicators can be calculated in this way. Overall, Figure 6.2 is a composite profile of an organization's relationships with its stakeholder community.

Such stakeholder information is vital to companies and stakeholders, who see the opportunities of sustainable enterprise. In the context of business ecology, such profiles as Figure 6.2 can help you see the relative strengths and weaknesses of your organization's stakeholder relations, and help you solve problems and develop new opportunities.

The organization shown in Figure 6.2 is excelling with shareholders and customers, but needs to strengthen relations with the environment, general public, communities, suppliers, and employees. The Malden Mills example in Chapter 4 demonstrates how maintaining these other relations produces innovation, quality products and services, and sustained earning streams. Well-treated, motivated employees and suppliers work harder and more consistently—and tend to be more innovative and cooperative. **Good public and community relations often pay dividends in lean times, such as the financial and logistics support given during the rebuilding of Malden Mills' textile plants.** A commitment to quality and environmental performance improves efficiency and often leads to new products, services, and markets. Considering that each of the plots shows value created for respective stakeholders, a profile of Malden Mills would be a larger polygon that more closely resembles a circle—indicating greater overall value creation that is more evenly, equitably distributed (see Figure 7.2).

Ralph Estes is leading a crusade to make information about corporate activities more accessible to all stakeholders—not just stockholders and corporate executives. In April 1996, Estes and his colleagues at the Center for the Advancement of Public Policy launched the Stakeholder Alliance program. Estes and the Stakeholder Alliance have been working with other groups such as the Interfaith Center on Corporate Responsibility, Co-op America, and the Jesse Noyes Smith Foundation, that have a similar interest in full and fair corporate disclosure and accountability, especially where community vitality, public health, natural environment, and general quality of life are threatened. Here are highlights of the program:

> Big business has great power with little accountability, and we know that power without accountability can lead to abuse. Abuse, in turn, provokes regulation, but regulatory efforts are often disappointing as agencies become captive to the very corporations they are charged with regulating.
>
> The Stakeholder Alliance seeks to bring about full and fair accountability by corporations to all stakeholders, to all those affected by the acts of corporations. We seek a level of corporate accountability commensurate with the level of corporate power.
>
> We propose a practical, two-stage approach. First, the public must come to recognize that corporations exist only through charters

granted by the people, and corporations therefore owe an accounting to us all. Once this requirement for public accountability is generally accepted, we can then reassert the public, social purpose for corporate charters.

We need full and fair disclosure of relevant information to all stakeholders, to enable customers, workers, host communities, neighborhoods, taxpayers, and others with a stake in the activities of large corporations to make rationally informed decisions and to protect their personal and economic interests. Fair disclosure will produce a stronger economic system while empowering the marketplace, the people, to discipline the corporate system and thus protect that system from its own excesses.

SUSTAINABILITY THROUGH STAKEHOLDER LEARNING

Sue Hall and Bryan Thomlison have worked independently and as a team to help companies integrate profitability, "stakeholder learning," and environmental performance. Their professional experiences with companies "leading the change" provide valuable insights into the promise of sustainable enterprise.

Sue Hall founded Strategic Environment Associates, located in Underwood, Washington, in 1992 to assist companies and other stakeholders in creating business-based solutions to environmental problems. She is also Executive Director of the Institute for Sustainable Technology (IST). Following are business profiles and excerpts from Hall's "Sustainable Partnerships," which appeared in the magazine *In Context* (Summer 1995).

As a research fellow at the Harvard Business School and a business consultant, Hall explored the following question: "Could companies gain competitive advantage from becoming environmental leaders?" Hall discovered that "Yes" was the answer, especially when companies co-create solutions with environmental stakeholders. Her work shows that **businesses are positioning themselves favorably within their markets by focusing time, energy, and money to develop sustainable solutions to environmental problems.** In some instances, companies are even changing the rules of their industry, forcing competitors and other players to change to more sustainable practices as well. According to Hall:

These proactive companies were more successful because they were willing to respond early to signals that the market was restructuring in response to environmental challenges and were prepared to partner, rather than oppose, environmental stakeholders to co-create solutions to tackle these issues.

Hall's research found that leading companies were improving their business environment by anticipating environmental and social forces that were reshaping the marketplace—including suppliers, retailers, customers, and other stakeholders. Hall refers to "stakeholder learning" as a valuable tool for gaining these competitive insights. The sustainable or "green" products created by this process are not only profitable, they actually propagate changes in the business environment. Hall provides a number of examples, some from the oil and gas industry.

Chevron, for example, has gained share in the oil exploration market by positioning itself as the "operator-of-choice" in environmentally sensitive regions. Chevron saw environmental concerns as an opportunity to differentiate themselves from their competitors. Meanwhile, according to Hall, oil is losing market share to natural gas because gas is cleaner burning. Apparently, traditional pollution "externalities" are being internalized within companies, particularly in areas of strict emission controls or where voluntary pollution prevention measures are considered cost-effective.

Henkel, one of Europe's largest chemical and detergent companies, positioned itself favorably by developing a phosphate-free detergent in the late 1970s. The company noticed the growing concerns over phosphate loading of rivers and streams in West Germany. Rather than ignoring or downplaying the problem, Henkel invested considerable research and development money to come up with a substitute—zeolite. By being the first company to introduce phosphate-free detergents, Henkel gained market share in Germany (from 16 to 23 percent) and France (gaining 6 percent). Parallel with this, Henkel joined forces with Degussa to capture a 70 percent market share of the zeolite production industry. Phosphate producers, their former competitors, suffered considerable losses from overcapacity.

Shaman, a California-based pharmaceutical company, is an example of a business who is learning from indigenous peoples to gain a competitive edge. Shaman conducted its "stakeholder learning" with traditional indigenous healers to develop drugs from native plants to treat various diseases. Remarkably, Shaman plans to share the profits from these drugs it develops with these stakeholders. According to Hall:

> When these plants are tested for effectiveness in treating those diseases, half the plants test positive—a hit rate over 50 times that of most drug companies.

Two Shaman drugs, now in phase II testing, may complete their Food and Drug Administration (FDA) trials within seven to eight years of their initial plant screening, compared to an average of 10-12 years of conventional drug companies. Since the FDA grants patent protection—and thus exclusive "monopoly" profits to drug companies for up to 17 years after initial screening, this could provide Shaman with up to four years' additional protected revenues and profits.

Shaman plans to share the profits from these potentially billion-dollar drugs with the indigenous communities from whom it first learned of such possibilities—offering an alternative revenue source to oil and timber extraction for these vulnerable peoples.

Hall found that while companies like Chevron, Henkel, and Shaman are seizing these opportunities, other companies are apparently ambivalent or unaware of these changes in their business environment. Hall compares the dramatic differences between choosing "stakeholder learning" or "denial":

> Companies have gained competitive advantage by leveraging the environmental forces that are already reshaping and restructuring their marketplaces to create profitable business solutions to environmental challenges. This success often depends upon the quality of learning and innovation that these companies build into their relationships with environmental stakeholders. As a result of these companies' leadership, whole markets have continued to restructure towards more sustainable solutions.
>
> By contrast, companies can choose to deny these new realities—and fail to introduce sustainable innovations. However, their businesses will continue to cause the environmental problems that fueled the market restructuring and ultimately create a downward, uncompetitive spiral. A profound lack of sustainability is certainly the hallmark of companies whose industries face "dinosaur extinction" status. Decommissioning costs for the nuclear industry, for example, and even Superfund clean-up costs for chemicals, place the long-term viability of both these industries in doubt unless ways can be found to pass the bill on to taxpayers.

Sue Hall describes the work of her colleague Bryan Thomlison when he worked at Arm & Hammer, the baking soda company based in New Jersey. Under Thomlison's leadership, Arm & Hammer discovered several new uses for baking soda, based on its nontoxic characteristics and versatility. These new uses produced a 30 percent increase in baking soda sales over a 36-month period. This increase was considerable, since the baking soda industry had been stagnant for decades. A key to these new discoveries were on-going discussions with stakeholders, including customers, environmental groups, government regulators, universities, and the media.

One example involves cleaning printed circuit boards. An environmental leader asked Thomlison if baking soda had ever been used to clean circuit boards. Traditional solvents used in the circuit board cleaning process were a major source of chlorofluorocarbons (CFCs) and volatile organic carbons (VOCs). Thomlison coordinated a study conducted by Arm & Hammer's research and development group and the environmental stakeholders. The result: two weeks later a prototype product was developed, which led to a new line of patented industrial cleaners.

Sue Hall describes the competitive advantages derived from "green" products and the "stakeholder learning" approach at Arm & Hammer:

> While 15 percent of company revenues are derived from the "green" market, the company's stakeholder approach alone contributes an entirely incremental 5 percent of revenues. These incremental sales are created by the additional "green" consumers that this uniquely powerful stakeholder approach attracts. Furthermore, Arm and Hammer has found that its stakeholder strategy is twice as cost effective as traditional marketing approaches, generating $10 for every $1 invested, compared to $4 for the company's traditional marketing approach— yet another source of competitive advantage.

Today, Bryan Thomlison has his own company in New Jersey, Thomlison Strategic Alliances, Inc. Like Hall, Thomlison is helping companies apply stakeholder learning to achieve higher levels of success. Thomlison's career has spanned twenty-five years and several well-known corporations, including Procter & Gamble, Carnation Company, Unilever, Church & Dwight (Arm & Hammer), and covered many aspects of business, including sales, marketing, public affairs, environmental management, and corporate philanthropy. More recently, he cofounded Our Community Development Corporation (OCDC), a Delaware-based business development company, which is "dedicated to the creation of businesses and jobs in inner-city America."

Thomlison sees business as a force for positive change on all levels and told us:

We believe that robust neighborhood economies cannot be re-established without the co-nurturing of business leaders. Each stakeholder sector offers unique skills and resources. Unfortunately, industry is conspicuous by its absence. Yet, a business presence is exactly what is needed for the sustainable economic health of a community.

Thomlison is putting his vision and ideas into action in Mercer County, New Jersey. Under the auspices of OCDC, Thomlison's innovative, stakeholder-driven business model will create up to 40 full-time and 22 part-time jobs in economically disadvantaged neighborhoods. Thomlison's project, called Our Community Enterprises, L.L.C. (OCE), is focusing on a green garment-cleaning business that uses alternatives to perchloroethylene, a hazardous chemical commonly used in the dry-cleaning industry. Thomlison believes alternatives to perchloroethylene offer strong growth potential. But, what makes the project particularly innovative is its innovative equity structure. Carl Frankel describes Thomlison's project in *Business Ethics* (November/December 1997):

> . . . Typically, when entrepreneurs set up in an economically distressed area, they create jobs but keep ownership for themselves. This structure often breeds distrust, especially when the owners are seen as outsiders.
>
> OCE is breaking with this tradition by bringing in the economically distressed West Ward community of Trenton as a one-third shareholder. The managing partners (Thomlison and two other entrepreneurs) also have a one-third share and investors will have another third. The businesses will be financed by Thomlison and his partners' own six-figure personal investment, plus $1 million being sought in start-up capital. If the business succeeds, it will be sold to high-achieving employees and local entrepreneurs (or community-based organizations). A local bank has already endorsed this 'set up-and-sell' concept and pledged to fund the contemplated employee buy-out. In this way, OCE will do more than create jobs. It will create local ownership as well.

Thomlison plans to replicate this process to help revitalize inner city communities across America.

Sue Hall and Bryan Thomlison are "change agents" in their own right. Their work has influenced corporate America by demonstrating that being socially responsible can mean both improved financial statements and improved communities.

PROFILES OF SUCCESSFUL COMMUNITY-BASED ORGANIZATIONS

Building community can make your organization more viable, profitable, and competitive, and create a healthier, more enjoyable work environment. Following are profiles of people and organizations that are creating sustained profit streams and quality-of-life success by building healthy relations with their employees, customers, suppliers, and other stakeholders. These include successful community-based organizations that are choosing and developing sustainable paths to the future.

Chaney Enterprises, Waldorf, Maryland, U.S.A.

Founded in 1962, Chaney Enterprises, a family-owned and operated company, is the leading supplier of concrete, sand and gravel, and concrete block in Southern Maryland. But this company is about more than profits. According to President Frank H. Chaney, II, his company:

> will never be a company that does not recognize its responsibility to the community. Wherever possible, we much reach out. A truly successful company must be a part of the community—an insider, not an outsider.

Chaney Enterprises has made employee retraining a top priority. According to Frank Chaney:

> Our employees are our most important asset. [Investing in them] projects a rewarding and healthy working environment. This in turn produces quality products, sound management, satisfied customers, and community growth.

In recent years, the company has undergone significant changes in leadership and business practices to meet the changing needs of its stakeholders, which include customers, associates, owners, families, and community. Says Chaney:

> The cultural climate is one based on trust and fairness to all—customer and associate. Our training programs reflect the corporate desire to improve the technical and interpersonal skills of each member of the Chaney Enterprises family. Our business practices require that we be

fair in our dealings with others, be honest and trustworthy with ourselves, and operate in a trusting environment with those we serve and lead.

Chaney, as chairman of the Maryland Aggregates Association, continues to lead change within the aggregates and allied industries, and to develop healthy relationships with state and local governments. He also advocates strong environmental standards:

> Maryland Aggregates Association, the largest association representing aggregate producers in Maryland, has worked diligently to establish and maintain a strong working relationship with state and local elected officials, public officials who are involved in the aggregates industry or use its products, and allied industries. In addition, the association is instrumental in the development of and compliance with specification, performance, reclamation, zoning, and environmental standards.

Chaney Enterprises participates in and sponsors community advisory groups in locations around Maryland that are near their production facilities to ensure that the needs and concerns of the stakeholder community are brought into business and operating decisions. With 34 years of business in Maryland and over 200 years of family history, the Chaney family has deep roots in the Washington, D.C./Baltimore, Maryland region and sees such activities as a responsibility of a good corporate citizen. Employees are encouraged and recognized for their participation in volunteer activities such as Boy Scouts and Girl Scouts, sports programs, and civic projects. Chaney Enterprises has also established a separate charity foundation, the Eugene Chaney Foundation, to assist community-based programs such as schools, hospitals, local youth recreation centers, and environmental and historical preservation organizations. They have also helped raise funds for national programs, such as the National Multiple Sclerosis Society and the American Kidney Foundation.

It is no surprise that Chaney Enterprises has received numerous awards for their commitment to ideals and excellence, including: the 1997 Reed W. McDonagh Business of the Year Award and the 1996 "Local Company of the Year" award from the National Multiple Sclerosis Society, and the Award of Excellence in the Manufacturing Facility category of the Associated Builders and Contractors, Inc. (Anne Arundel/Southern Maryland Chapter) in 1992 and 1993.

Chaney Enterprises sums up their philosophy in the following corporate statement:

We are committed to improving the quality of life for our customers, associates and the families of our community through quality products and services delivered in an ethical manner. We are building a firm foundation for Southern Maryland.

The Company's associates, more than 300 strong, produce and deliver sand, gravel, concrete, block, and concrete related supplies to a variety of customers throughout Southern and Eastern Maryland. Chaney Enterprises maintains significant reserves of land, from which we produce our primary products, and develops each tract with consideration for the needs of the Company, the local business and residential community, and the natural environment.

John Bichelmeyer Meat Company, Kansas City, Kansas, U.S.A.

Dick Sumpter works in Kansas City at the U.S. Environmental Protection Agency (Region 7) and conducts workshops on strategic planning, sustainable development, and organizational change. For the Business Ecology Network's newsletter, *Main Street Journal,* Dick shared his "wisdom from the heartland"—as learned from his father-in-law, John Bichelmeyer. This common-sense approach to management shows how a company with a clearly defined purpose and a strong set of values, can achieve and sustain higher levels of success, as Sumpter explains:

"America's Heartland" usually refers to the broad central expanse of rolling hills and prairie known as the Midwest. Much of the economy is rooted in agriculture. The world's largest cooperative, Farmland Industries, is headquartered in Kansas City, and the states of Iowa, Kansas, Missouri, and Nebraska contain 122 million acres under cultivation. While there are a number of large corporate farms, much of the acreage is still family owned and farmed. These folks, of necessity, are conservationists. They have to be if they want to pass on this way of life to their children. But the pundits on either coast misread this and insist on labeling them "conservative."

One of the people who epitomizes this heartland philosophy is my father-in-law, John Bichelmeyer. He is 80 years old now, but still very active, mentally as well as physically. He grew up in a cabin with a dirt floor near Eudora, Kansas. He trapped and hunted for food, and at an early age learned that while we may call nature "Mother," she can be anything but kind. When game was scarce, he would pick potatoes for a neighbor to help feed the family. During the Depression, at the age of 14, he left high school to work in a meat packing plant. But in 1946, he opened his own meat processing plant and market not far from the plants he worked in.

That business will be 50 years old this year. It proved to be a wonderful institution of higher learning for those who worked and traded there. John was the professor-in-residence, and he used this place to teach his four sons who worked there many valuable lessons. (He was fond of saying that "to make it in this business you have to sell everything but the squeal." There was no waste permitted.) When I came along decades later, I was also an eager student of his. On one occasion, when John was waiting on customers, a young consultant asked if he could speak with him. Gesturing to the counter, full of customers, John asked him to come back another time. The consultant, who was an "efficiency expert," worked mainly for the large packing houses and thought to peddle some of his insights to my father-in-law. He told John, "Now we both know why you're in business—to make money."

"That's your first mistake, son," said John. "I'm not in business to make money, I'm in business to make a living. I don't need you to tell me that I could make more money if I stayed open until six o'clock; or if I was open six days a week instead of five. Tell me, son," he went on, "what else do you do for a living?

Why don't you have another job? Never mind, I'll tell you. Because some things are more important than money. Time for your family, your community. I've had many opportunities to get bigger, to expand to the next level. But frankly, I don't like what I see up there. Those folks are in business to make money. I only want to make a living."

With that philosophy John did make a living. He educated ten children, some in his own schoolhouse and some in more traditional settings. He put all six daughters through college, two with masters degrees and one with a doctorate. Two of the sons graduated from college, and two others left after a year or two, feeling that they learned more at "Bichelmeyer U." And I think they made the right choice. I've never met people with a greater sense of business integrity than this group of father and sons. They know the true meaning of Quality, and to my knowledge, never had a course in TQM. The business has flourished for half a century with virtually no advertising; a testimony to the value of satisfied customers.

Clearly, Bichelmeyer's business values include common sense, a strong sense of family and community, being attentive to his customers, integrity, quality products and services, and finding a balance between work and private life. Like any small family-owned business, the John Bichelmeyer Meat Company has not been without hard times. After just two years of operation, a 1908 flood destroyed the original Bichelmeyer market, run by John's father George Bichelmeyer. A bigger flood in 1951 wiped out Bichelmeyer Meats in only its fifth year. Friends and employees worked without pay to help the family rebuild in 1951. A fire on April 5, 1995, completely destroyed the building. Once more, the Bichelmeyers rebuilt on the same site. Bichelmeyer credits their longevity and resilience to dedicated employees and relationships; he built loyal bonds with customers and a community that in turn saw him through tough times.

Bichelmeyers's kernel of wisdom to aspiring entrepreneurs? "The first hunk of meat you sell is yourself."

The World Bank, Washington, D.C., U.S.A.

Spirituality, an environmental ethic, and civic-mindedness are changing business culture. Corporate leaders are recognizing their roles as instruments of positive change and are actively participating in their communities and taking responsibility for economic, environmental, and social issues.

For example, the World Bank sponsored back-to-back conferences in Washington, D.C., in 1995 that addressed spiritual values, ethics, and financing related to sustainable development. The late Willis Harmon, co-founder of the World Business Academy and the Institute for Noetic Sciences, was one of the people who shaped this event. Harmon clearly saw the enormously important role business can and should have in creating a sustainable future when he wrote:

> More and more people are coming to believe that the modern world is undergoing a period of fundamental transformation, the extent and meaning of which we who are living through it are only beginning to grasp. Fewer people seem to realize that in that transformation, the role of business, because of its central place in modern society, is absolutely crucial. It will either be a major facilitator of that change, or it will be the greatest obstacle.

Harmon's considerable leadership and vision shaped us when he was alive. He is missed by many, but his spirit will guide us forward to achieve the transformations that he and other leaders have started.

The openness, heightened consciousness, and sense of mission exhibited by the diverse group of business, environmental, and spiritual leaders present at these conferences was truly remarkable. Many participants and organizers, such as Richard Barrett of the World Bank and

John Hoyt of the Center for Respect of Life and Environment, have already created significant improvements where they work and live. This includes innovative empowerment programs for citizens and communities, reinventing powerful institutions though internal reform and activism, and leveraging huge financial and institutional resources to support sustainable development.

Richard Barrett, for instance, founded the World Bank Spiritual Unfoldment Society in March 1993 to provide a forum where Bank staff could discuss and exchange beliefs and ideas that promote spiritual awareness, and integrate this awareness into their personal and professional lives. This Society of several hundred people continues to hold retreats, workshops, and weekly meetings to create within the World Bank a conscious understanding and caring that transforms the way staff work with each other and their contacts around the world.

Food from the Hood, Los Angeles, California, U.S.A.

A marvelous example of unleashing entrepreneurial spirit and ingenuity to revitalize communities is the "Food from the Hood" project in Los Angeles, California. This highly successful project sprouted literally from the ground up. Following the devastating L.A. riots of 1993, a group of committed citizens, which included mostly youth and young adults, planted a community garden to help renew civic pride and foster self- reliance. This community-based enterprise started with growing and selling vegetables and herbs, and eventually developed and sold salad dressings to local farmers markets and grocery stores. Impressed with the entrepreneurial spirit of the neighborhood, local business leaders began providing financial and technical assistance. Eventually, the salad dressings were distributed regionally through several chain and specialty stores. A foundation was formed from the proceeds to help support training and college scholarships for neighborhood kids.

Vermont Businesses for Social Responsibility, Vermont, U.S.A.

In 1995, Vermont Businesses for Social Responsibility helped pass state legislation supporting the development of jobs that have positive impacts on the environment and communities. Here is how Jane Campbell and Pat Heffernan, Vermont business leaders who helped develop the legislation, reported it to *In Business* magazine (September/October 1995):

At the core of the Sustainable Job Fund (SJF) is recognition of a daily reality for small business owners—they are often at a financial disadvantage, lacking economies of scale of larger companies. Furthermore, for businesses with above average work place quality or environmental practices, the extra costs of being "good" puts them at an even greater price disadvantage in the marketplace.

SJF will help small businesses work together on job creating projects which they couldn't individually afford to do, such as collaborative marketing, research, or production. Projects to be given preference involve environmental technologies; energy efficiency or pollution prevention; natural resource-based industries such as specialty foods and forest-based products; or activities that will benefit the quality of Vermont communities. Two examples of possible SJF projects might be: A group of dairy product manufacturers working at the university to turn dairy waste into new marketable products; or a regional group of businesses forming an employee child care center that's cooperatively owned by the businesses.

The SJF is also striving to help other states, business groups, and regions develop similar programs.

Edible Schoolyard, Berkeley, California, U.S.A.

Alice Waters, the founder of Chez Panisse Restaurant in Berkeley, California and an organic food activist committed to sustainable agriculture, was cited in Chapter 4. In a letter she wrote to President Bill Clinton in December 1996, she describes the Edible Schoolyard as an "on-the-ground" community revitalization program that brings together key stakeholders: businesses, schools, students, teachers, and community groups. She started the program at the Martin Luther King Jr. Middle School in Berkeley, California, and hopes to propagate it across the U.S. Says Waters in her letter:

> But we need much more than the physical renovation of the schools. We need to renovate the meaning of education itself. The aim of education should be to provide children with a sense of purpose and possibility, and with the skills and habits of thinking that will help them to live in the world, whether or not they go to college. I believe that the one key thing children can learn that teaches these skills and habits is how to eat well and rightly. A curriculum which educates both the senses and the conscience—a curriculum based on sustainable agriculture—can inspire children by teaching them a moral obligation to be caretakers and stewards of the finite resources of our planet.

This is why we envisioned the program known as the Edible Schoolyard here in Berkeley at Martin Luther King Jr. Middle School. Our mission at King School has been to create and maintain an organic garden and landscape that is integrated with the school's academic curriculum and lunch program and that involves students in all aspects of farming the garden—as well as in preparing, serving, and eating the food they grow. This will stimulate and educate their senses and teach them the transformational values of community responsibility, good nourishment, and good stewardship of the land.

The students have helped create the garden through active partnerships with teachers, neighbors, and businesses in the community, who have all volunteered goods, services, and time to transform blacktop into topsoil and to restore buildings and grounds. These students are hungry, physically and spiritually: for real food, and for role models. Thanks to our volunteers, many of whom are young people from Americorps, the students have been made to feel that there are people who care about them and who want to make their school a beautiful place.

Colegio de Estudios Superiores de Administración, Colombia

In Colombia, Colegio de Estudios Superiores de Administración (CESA) is cultivating a new breed of business leaders who use "principled private initiative" to revitalize communities. Assuming responsibility for "giving business leadership capabilities to the community," a group of dedicated businessmen founded CESA in 1974 to blend pragmatic enterprise with purpose and high moral character. Located on the edge of Bogotá in the foothills of the Andes, this incubator of socially responsible business leaders has educated about 1,200 Colombians and international students over the past two decades.

This school of higher learning creates opportunities for community-oriented students. Marco Fidel Rocha, CESA's current director, remarks: "We are looking for young people of high moral character and awareness, who love Colombia and want to create a better community through economic opportunity." The school is credited with creating 10,000 new jobs and with 100 percent of its graduates being employed. Most have started their own businesses.

Beatriz and Eduardo Maciá, two of their more successful graduates, were highlighted by David Benson in the August/June 1996 issue of *Fast Company* magazine:

Eduardo Macía, 37, gets a little heated when asked to discuss the financial performance of the restaurant business he and his wife, Beatriz, 36, founded 16 years ago in Bogotá, Colombia. It's not that the numbers aren't impressive: the Macías, sole owners of a chain of 14 Crepes and Waffles restaurants throughout Colombia, Ecuador and Panama, report a growth rate of 30 percent for each of the last five years; the company now employs 700 people.

What Macía objects to is the classic definition of success only in terms of financial gain. "Life is not only money," he insists. "If you do things right and take care of your employees and customers, the profits will come by themselves."

For the Macías, doing things right involves helping finance more than 80 homes for their employees, providing free health insurance after only one year of service, and having a social worker available for home consultations. Beatriz Macía explains, "To be a leader is not just to build a business, but rather to contribute to the country's development. To raise the quality of life in every way."

Companies like the Macías' are reaching higher levels of success—including sustained earning streams—by building strong relations with their employees, customers, suppliers, communities, and other stakeholders. Like organisms living within ecological communities, businesses both affect, and are affected by, what happens within their web of relationships and communities. Business ecology is a powerful lens for seeing the framework of these vital relationships and rejuvenating them to create sustainable, life-sustaining enterprises.

In the early sections of this chapter, we discussed why aligning vision, values, and purpose and building strong stakeholder relationships are critical to organizational success. As Ralph Estes emphasizes, company managers are motivated under the present system to maximize the bottom line for shareholders, often at the expense of nonshareholder stakeholders who have vested interests in a company. **Developing a sustainable enterprise requires a new measuring system—one that articulates shared values, vision, purpose, and key relationships. This is essential to building connections and community both inside and outside your organization.** Systemic profiles, such as Figures 6.1 and 6.2, can help your organization assess its strengths and weaknesses and develop strategies for becoming a more viable, sustainable enterprise.

Learning from stakeholders, as demonstrated by Sue Hall and Bryan Thomlison, can be an effective strategy for developing sustainable enterprise. In fact, successful companies, such as Henkel, Shaman, and Arm &

Hammer, demonstrate that profitability, social responsibility, and environmental performance can be integrated for competitive advantage.

We also learned from other organizations, such as John Bichelmeyer Meat Company, which has endured the ups and downs of business by adhering to values that build community and enduring stakeholder relations. Companies like Chaney Enterprises and Chez Panisse Restaurant see giving back to their communities to be a natural part of doing business. Finally, groups such as Vermont Businesses for Social Responsibility are choosing and developing sustainable paths to the future.

Leaders and managers who understand and develop healthy stakeholder relations will be on the cutting edge of the emerging ecological economy. Business ecology puts "community" back in business by being relationship-oriented and encouraging cyclical flows that reinvest goods, services, and profits into the local economies that created them. It includes all stakeholders, renews vital relationships and flows, and creates an organizing framework modeled on natural systems.

7

The Shift to Sustainability

A seed hidden in the heart of an apple is an orchard invisible.
Welsh proverb

Sustainable development and ecological restoration are inevitable. The message is clear: Life is the real bottom line and those organizations able to grasp the implications of sustainable development, embrace its values, and transform themselves into sustainable enterprises will have a clear, competitive advantage in the next economy. Learning and applying business ecology is the most effective means for transforming your business or organization into a life-sustaining enterprise that thrives in the emerging ecological economy. Business ecology synthesizes a holistic view of the "bottom line" to integrate profitability, values-based management, stakeholder relations, life-cycle thinking, and environmental performance, giving your organization a "natural edge" in today's rapidly evolving business environment. In Chapter 1, we introduced the seven seeds of business ecology. Chapters 2 through 6 outlined in detail business ecology's mindware for the new millennium, models and strategies that emulate natural systems and provide an elegant, relationship-oriented approach to managing business and organizations.

This final chapter helps you grow business ecology within your organization. Following are effective strategies for cultivating the seeds of business ecology, including how visionary leaders and organizations such as Monsanto's Robert Shapiro and Interface's Ray C. Anderson are making a fundamental values shift—to transition into sustainable enterprise. To help your organization be successful with making such a shift, consider joining our nonprofit learning community—the Business Ecology Network (see Appendix).

1.   CREATE A VALUES-BASED ORGANIZATION

Your organization, as a living organism, is most successful when its development and behavior are aligned with its vision, values, and purpose—its "social DNA"—and its environment. Business ecology, as a values-based, organic model for organizations, creates an inner core of community, continuity, and resiliency. This social DNA which defines your organization's identity, also selectively filters value-creating flows and relationships that sustain it within its environment. These flows and relationships, in turn, define your organization's internal economy (metabolism), how it makes a living (niche), and where it lives (habitat).

Highly effective people and organizations are values-driven. Values give them a well-defined identity and a clear sense of purpose. This self-knowledge helps individuals, businesses, and organizations weather uncertainty and rapid change in their environment by providing a fixed reference, much like stars do for navigators. Core purpose and values are the social DNA, the genetic codes of human systems, shaping how these systems organize and develop. It is shared core values which define the "self" in self-organizing human systems.

Creating a values-based business or organization requires a step by step assessment of core purpose and values and open communication of this vision with all stakeholders. Xerox Business Services and Malden Mills, described in earlier chapters, are good examples of businesses that are adaptive, successful, and values-driven. Chris Turner of Xerox Business Services describes how values, or principles, link with a shared vision and open communication in her interview with *Fast Company* magazine (October/November 1996):

> We knew the first piece of our strategy—to create a shared vision. But I never thought of it as a written vision statement. To me, a vision is an ongoing conversation. It's the way we think, individually and collectively, about the community we're creating. It's the principles of the people in the organization. What's important to us. How we want to be with each other. It's never frozen, it's never set. It's energy—or spirit.

How do you create a values-based organization? James Collins and Jerry Porras, authors of *Built to Last,* describe the importance of linking an organization's core values, purpose, goals, and vision into a visionary framework. Core values are the qualities that define an organization's identity. The purpose is why the organization exists beyond the notion of profitability. For example, the core purpose of Merck, the pharmaceuticals company, is "To preserve and improve human life." Core values and purpose together form the core ideology of an organization, defining "what it stands for" and "why it exists." A vivid description of a desired future, coupled with bold but achievable goals, complete the visionary framework. Collins and Porras describe a healthy, dynamic tension—"preserving the core (ideology)" while "stimulating progress"—that exists within visionary companies. Creating a vision and goals challenges and excites people within an organization, while the core ideology provides the genetic blueprint, the cohesive framework, for making the vision and goals come to life.

Collins and Porras emphasize that core values are discovered, not created, by looking within an organization to see what it really stands for. This is what the business ecology lens can help you see: the true values-center of your organization. Business ecology answers the question: What are the qualities that actually describe the culture of the workplace? Interestingly, Collins and Porras allow for the revitalization of core values that have weakened over time, but contend that values that describe how the organization "ought to be" do not belong as core values. Their reasoning is that people within the organization will become cynical of the whole process, because these values are not authentic representations of the organization. Collins and Porras also emphasize that the values are internally driven, and do not necessarily have to be meaningful to outsiders. In fact, they go on to say that values often determine who does and does not belong to an organization.

While Collins and Porras provide many insights into values-based organizations, it is important to highlight why and how business ecology differs from, and is consistent with, their framework. Like the Collins and Porras framework, business ecology recognizes that values guide and motivate people within an organization. Collins and Porras describe how values either attract or repel people to an organization. **With business ecology, you can create organizational change by defining core values,** even new core values that are shared by individuals within an organization. **This inner integrity creates outer resiliency, flexibility, and the strength to filter new ideas and information continually—in other words, to keep adapting as needed.** Just as in life, these values can adapt to transition an organization for desired growth and change. Consider the change in values that has taken place within companies such as Interface, Monsanto, Mitsubishi, and Dupont—how else could they change to embrace a path to sustainability? A values shift is often critical for businesses and organizations which seek genuine systemic change. Business ecology takes this extra step of not only defining current core purpose and values but setting goals for a values shift to sustainability—and how to transition to meet such a goal.

Business ecology recognizes that values, as social DNA, shape a business or organization's "membrane," determining what life-sustaining flows are filtered in and out of an organization. These life-sustaining flows, described in detail in Chapter 4, include: people and other organisms; food, information and ideas; money, products, and services; air, energy, and water; and materials. The values of an organization, in fact, determine what and who is valued. A bottom-line-driven organization, for example, often misses opportunities for improving ecological efficiency and viability by focusing exclusively on cash flow.

Business ecology also makes the distinction that sustainable businesses and organizations are not insular, but rather, they are part of a larger system of relationships and need to take their stakeholders into consideration, whether it be the local community or regional suppliers, when identifying core values. In fact, businesses and organizations that emulate the organizational design of successful natural systems are both self-referring, in that they have clearly defined, shared values, and are part of a larger web of relationships with varying types of stakeholders and organizations—even competitors. Strong core values and attention to building relationships will ensure success.

While being aligned with organizational values is an effective strategy for long-term success and weathering change, it is important that an organization be open as well to flows from the outside. Organizations that remain too isolated, or cult-like, may loose touch with changes in their environment. This is particularly risky in a period of major discontinuity and change, such as the period in which we presently live. Being alert and ready for change is essential.

Importantly, business ecology also recognizes that social DNA, core purpose and values, is one of the types of information flow exchanged among organizations. Like the mixing of DNA in natural systems, organizations share social DNA to keep evolving. Leaders and stakeholders carry social DNA and in effect "cross pollinate" new organizations that they join. This mixing of social DNA, this being open to new ideas and values, enables businesses and organizations to select for the strength and adaptability needed to transition into sustainable enterprises and to be competitively positioned in the new ecological economy. Business ecology provides a framework, similar to that of Collins and Porras, to the extent that social DNA shapes an organization in every way (actions, goals, design, culture, etc.). However, business ecology is different in that it recognizes that proposed values of "what we ought to be" bear equal weight to existing or previously held but waning organizational values. In fact, **an infusion of sustainability values—a values shift—is precisely what most organizations need today.**

A truly successful "organism" is values-based and interacts in a healthy manner with its environment. Business ecology, by helping businesses and organizations align with their social DNA—their core purpose and values—and their environment, is a powerful catalyst for a sustainable enterprise.

 ## 2. ADAPT AND THRIVE

Leading companies are already applying systemic, closed-loop thinking to create enduring success and competitive advantage. Your business or organization can emulate these visionary companies by applying business ecology's organic model to create a dynamic learning organization, strategically positioned with vision, foresight, self-knowledge, and systems thinking. This model can help your organization identify and maintain its core identity while remaining open to the changing environment. Your business or organization can learn to adapt, innovate, survive—even thrive—in the ecological economy.

Stuart L. Hart, former director of the Corporate Environmental Management Program at the University of Michigan Business School and currently associate professor of management at the University of North Carolina, Chapel Hill, describes both the challenge and opportunity posed by sustainability in "Strategies for a Sustainable World" in *Harvard Business Review* magazine (January-February 1997):

> It is easy to state the case in the negative: faced with impoverished customers, degraded environments, failing political systems, and unraveling societies, it will be increasingly difficult for corporations to do business. But the positive case is even more powerful. The more we learn about the challenges of sustainability, the clearer it is that we are poised at the threshold of a historic moment in which many of the world's industries may be transformed.
>
> To date, the business logic for greening has been largely operational or technical: bottom-up pollution-prevention programs have saved companies billions of dollars. However, few executives realize that environmental opportunities might actually become a major source of revenue growth. Greening has been framed in terms of risk-reduction,

reenginnering, or cost cutting. Rarely is greening linked to strategy or technology development, and as a result, most companies fail to recognize opportunities of potentially staggering proportions.

How can you transform your organization into a sustainable enterprise? Business ecology spotlights those elements of organizational design, such as values, perspective, behavior, and even culture, that are essential. It shows the key relationships that link these intangible design elements to concrete systemic results, including profitability, stakeholder relations, and environmental performance. Business ecology's systemic lens allows you to see simply and systemically how your organization really works within its environment.

Global economic, social, and environmental trends are rapidly changing the business environment. Is your business or organization ready? What is your organization's perspective? Its organizing framework? Its values? **Organizational transformation requires a simultaneous shift in values, perspective, and behavior** (see Figure 2.1). Values, as social DNA, shape the lens through which we see the world, and are the basis for decisions, actions, and organizational development. Perspective, as the lens through which we see and interpret reality, in turn, affects how our values develop and how we behave. Behavior is how we act and relate within our environment. Environmental feedback regarding our behavior shapes our values and our perspective.

Business and Sustainable Development

How are businesses responding to sustainable development? **Several prominent business leaders, such as Robert B. Shapiro, chairman and CEO of Monsanto, Ray C. Anderson, chairman and CEO of Interface, Inc., and Tachi Kiuchi, managing director of Mitsubishi Electric Corporation, see sustainability as a major discontinuity affecting their organizations and society as a whole.** Our economic well-being and that of business, is tied to a healthy global environment and society. Recognizing both its challenge and opportunity, leaders like Shapiro, Anderson, and Kiuchi are urging their organizations to evolve rapidly into sustainable enterprise. In the context of business ecology, this adaptation to sustainability raises two important questions: **What are the values of sustainability? Are your organization's core values consistent with the values of sustainability?**

We introduced sustainable development, or sustainability, in Chapter 1 as the successful integration of economic, social, and environmental goals to ensure the quality of life today and in the future. It is a long-term, values-based process that involves all stakeholders. It is about having a vision, and choosing, planning, and creating a human society that recognizes our interdependence with the broader ecological system of the planet. Sustainable development is also about equitable distribution of resources and opportunities. It encourages both interdependence and self-reliance, and balances competition with cooperation, to create more viable businesses, communities, and regions. The values of sustainability include time-tested natural principles, such as adaptation, feedback, life cycles, collaborative networks, and whole system efficiency, and the enduring wisdom of human cultures which preserves community, connection to place, stewardship, and spirituality. Table 2.1 illustrates how these values of the new ecological economy compare to the dominant values of the industrial economy.

Companies such as Monsanto and Dupont, once considered adversaries of the environment, are leading the change to sustainability among larger companies. Dupont, for example, has adopted a "zero emissions" goal for its operations, and is working with its stakeholder community to improve its overall ecological efficiency.

Under Robert Shapiro's leadership, Monsanto is rapidly evolving to realize lucrative opportunities with sustainable enterprise. Seven "sustainability" teams were formed to help the company strategically position for competitive advantage. The following, which is put in the context of business ecology, is derived from Joan Magretta's interview of Shapiro in *Harvard Business Review* in January-February 1997:

1. **The Eco-Efficiency Team:** This team measures the ecological efficiency of Monsanto's processes and elevates the importance of life-sustaining flows, such as energy and water, that have been undervalued or taken for granted within the traditional accounting framework. This work is laying the groundwork for a better understanding of metabolism within and among the company, its suppliers, customers, and other stakeholders.
2. **The Full-Cost Accounting Team:** This team is developing a methodology to account for the total cost of making and using a product during its life cycle, including its production, use, recycling, and disposition. The goal is to make more informed decisions about the true costs associated with each product's life cycle.
3. **The Index Team:** This team is developing indicators or criteria for evaluating how different business units are moving toward sustain-

ability, that will be linked to Monsanto's broader management score-card. Indicators being developed by this team will integrate economic, social, and environmental factors related to products, services, and entire enterprises. The management scorecard, which includes finan-cial targets, customer satisfaction, internal processes, organizational learning, and sustainability indicators, fits well with the business ecol-ogy model. It explicitly acknowledges other vital flows and systemic relations that sustain Monsanto. Later in this chapter, several systemic profiles are presented related to values, stakeholder relations, and or-ganization viability.

4. **The New Business/New Products Team:** This team is exploring what products and services will be in demand in a sustainable society. This includes evaluating the technologies and skills that Monsanto can ap-ply to reduce environmental degradation and stress, prevent pollu-tion, and even restore ecological systems. In the context of business ecology, this team is anticipating dramatic shifts in Monsanto's busi-ness environment. That is, its new line of products and services will reflect the core purpose and values—the social DNA—of sustain-ability.

5. **The Water Team:** This team sees the quantity and quality of water resources, and the provision of related services such as drinking water, irrigation, and wastewater treatment, as critical global issues. Clearly, like business ecology, Monsanto sees water as a vital flow—the liquid asset—supporting life and human civilization. Monsanto is examin-ing how their expertise and technologies can solve worsening global water problems.

6. **The Global Hunger Team:** This team is focusing on another life-sustaining flow—food. Because Monsanto's roots are in agriculture, the company is exploring how this expertise and modern applications, such as biotechnology, can help alleviate world hunger.

7. **The Communication and Education Team:** This team is developing sustainability training to give Monsanto's workforce of 29,000 em-ployees a common perspective and identity. Information flow both within and outside the company is supporting this sustainability edu-cation program.

It is inspiring to see Monsanto, Dupont, and other larger companies adapt their operations and shift their values toward sustainability. In addi-tion, a growing number of smaller companies are already predisposed to sustainable enterprise. One is Sustainable Technologies Corporation (STC), a small "ecopreneurial" firm based in Lewisburg, Ohio. It embodies the thinking of business ecology and sustainable enterprise, shaped largely from the values of its founders, Mike and Julia Castle. The Castles are mindful of the waste and ecological destruction associated with our current industrial economy, and have developed a series of sustainable technologies for recycling residual material flows from paper products and agriculture. STC's product, Macrex®, for instance, is a wood-like, fire-retardant material that is made primarily from secondary paper-waste fibers. It can be milled like wood and altered to resemble wood and other materials. Sustainable Technologies Corporation is working with a British firm to develop a number of applications for Macrex®, including manhole covers, landscaping timbers, and building applications. This product line closes the loop on the enormous flow of secondary paper while offsetting the need to cut virgin timber.

PulVerdé®, another STC product, is a pseudo soil, a growth medium composed of manipulated cellulose fiber and nutrient amendments that is intended to replace top soils, potting soils, and peat moss, all materials that are often mined and developed in ways that are ecologically destructive. The PulVerdé® pseudo soil is 100 percent organic certified, and is being marketed to landscaping firms and greenhouse operations.

Blow-up, Inc., a professional color lab in Baltimore, Maryland, is another small company dedicated to closing the loop, especially when it is economically viable. The company recovers silver and other heavy metals from its photographic development process, collects and reuses plastic slide and film boxes, recycles batteries, reuses packing materials, such as envelopes, cardboard, and Styrofoam (e.g., packing peanuts), and participates in curb-side recycling of paper, cardboard, plastic, and glass. According to vice president Michael Foreman, Blow-Up, Inc. pursues cost-effective measures first, then looks for low-cost strategies for recovering materials that are less valuable or bulky, such as film cuttings from negatives. Clearly, collaboration among color labs and economic incentives from film producers could make recycling of bulkier or lower value materials more cost effective.

Another sustainable enterprise is Nugent's Upholstery, in Gulfport, Mississippi. Sue Nugent, the owner, saw opportunities for reusing mountains of discarded tires as a framework for her furniture. Becky Gillette describes Nugent's Upholstery in "Turning Tires Into Furniture," which appeared in *In Business* magazine (January/February 1997):

"All you have to do is drive around, and you'll see old tires lying on the sides of roads. Then, a local company that had been taking old tires and receiving disposal fees was found to have illegally dumped 300,000 of them."

Nugent, the owner of Nugent's Upholstery and a company called Your Environmental Solutions (YES) in Gulfport, Mississippi, decided to see if she could do something about the situation. "I've always been big on recycling," she says. "My upholstery company essentially is a recycling business. I keep furniture out of landfills." She started experimenting with different designs that use the old tires as the framework for furniture. Although the first pieces were not sturdy enough, she learned how to improve the design through trial and error. Besides finding a new use for old tires, her furniture greatly reduces the amount of wood needed for the framework.

According to Nugent, Mississippi alone discards about 2.5 million tires each year. Closing the loop with tires, wood, and other residual flows offers enormous opportunities for profit that are also good for the environment and our communities. Nugent hopes to spawn similar upholstery businesses across the U.S. Large and small companies such as Monsanto, Dupont, Sustainable Technologies Corporation, Blow-Up, Inc. and Nugent's Upholstery, represent a new generation of businesses that are evolving into sustainable enterprises.

Business ecology supports sustainable enterprise in many ways. It can help your organization (1) develop cyclical, life-sustaining flows in its internal and external economy; (2) align economic, social, and environmental goals; (3) meet "real needs" more elegantly by creating high quality, appropriate products and services from the efficient metabolism of life-sustaining flows; and (4) create and realize new business opportunities related to sustainable development.

3. USE A NEW LANGUAGE FOR SUCCESS

Business ecology is the language of sustainable enterprise and the ecological economy. This new language articulates the transition from industrial, mechanistic thinking to biological metaphors for management and design. It describes holistically your business or organization's well-being; aligns its vision, values, and purpose, defines its value creation relative to all of its stakeholders, and enhances the viability of your business or organization's life-sustaining flows and relationships within its environment. As a sustainable enterprise, your organization's success is defined by how effi-ciently it optimizes value for stakeholders while enhancing and sustaining life. In this sense, its value creation, and ultimately its success, is measured by the extent to which quality of life genuinely improves.

Organizational health is often summarized by bottom-line numbers. While the balance sheet, cash flow, and income statements are essential to understanding financial health, they do not account fully for an organization's well-being, such as its purpose and values, use of resources, creativity, employee satisfaction, and customer loyalty. We need a new language for success that meets today's challenges. Business ecology takes the first step of defining a common language that links the past with the future.

This new language of sustainable enterprise encompasses all the dimensions of an organization's well-being, including identifying and applying its core purpose and values, its value creation relative to all of its stakeholders, and the viability of life-sustaining flows and relationships within its environment. **A sustainable enterprise's success is defined by how efficiently it uses life-sustaining flows, time, and space to optimize value for stakeholders while enhancing and sustaining life. In this sense, its value creation, and ultimately its success, is measured by the extent to which quality of life genuinely improves.**

Many people and organizations are rediscovering genuine, enduring success by focusing on values and what brings meaning and purpose to life. How did we drift from a deeper, soulful understanding of "success" to one that is superficial and externally driven? Certainly, the commercial, entertainment, and consumption society in which we live cultivates and rewards more superficial, materialistic forms of success. In terms of motivating and inspiring people, it is important to distinguish superficial images of success from more enduring, genuine success.

Stephen Covey, author of several books and a renowned expert in purposeful living and leadership, makes an important distinction between what he calls the "Character Ethic" and the "Personality Ethic" in *The Seven Habits of Highly Effective People.* After studying about 200 years of U.S. literature, Covey discovered that the last fifty years defined success in terms of "image" or the "Personality Ethic." According to Covey: "Success became more a function of personality, of public image, of attitudes and behaviors, skills and techniques, that lubricate the processes of human interaction." In contrast, the first 150 years tended to focus on the "Character Ethic," which Covey defines as "basic principles of effective living," and the belief that people can only experience true success and enduring happiness as they learn and integrate these principles into their basic character. Covey describes the "Character Ethic" as a primary level of success, but the "Personality Ethic" is a secondary, useful means for attaining the primary principle-based success (i.e., the "Character Ethic"):

> I am not suggesting that elements of the Personality Ethic—personal growth, communication skill training, and education in the field of influence strategies and positive thinking—are not beneficial, in fact sometimes essential to success. I believe they are. But these are secondary, not primary traits. Perhaps, in utilizing our human capacity to build on the foundation of generations before us, we have inadvertently become so focused on our own building that we have forgotten the foundation that holds it up; or in reaping for so long where we have not sown, perhaps we have forgotten the need to sow.

This is similar to business ecology's distinction between actual and apparent value creation. Why make this distinction? Because one, actual value creation genuinely improves quality of life for customers and other stakeholders through high-quality products and services; enduring, rewarding jobs; a stronger, more vibrant economy; and other recognizable forms of value created by an organization. The other, apparent value creation, can be superficial. It is the quality of life improvement that we believe or expect to receive from a product, service, or organization. It may be greater, lesser than, or equal to actual value creation. It may be illusory. In our current society, apparent value creation is often greater than the actual value creation and shapes marketplace behavior. This produces overconsumption in a world of limited resources and growing poverty, while the genuine quality of life in our consumer-based societies actually erodes from the misallocation of time, income, life energy, and other life-sustaining flows.

Figure 7.1, adapted from *Your Money* or *Your Life*, shows the relationships between quality of life and material wealth. In our consumption-driven economy, overconsumption is often cultivated by the allure of apparent value creation. It often stimulates consumers to buy products and services that they do not need. Generally, the actual quality of life improvement experienced through many products and services is often less than what customers expect to receive. The consumption-driven marketplace convinces many people that they can never have enough, and pushes them to the downward side of the curve, where they have more than they need and are overburdened with taking care of their possessions and still trying to acquire more and more. The result: with each increase in material wealth their quality of life is diminished, not improved. Further, many people on

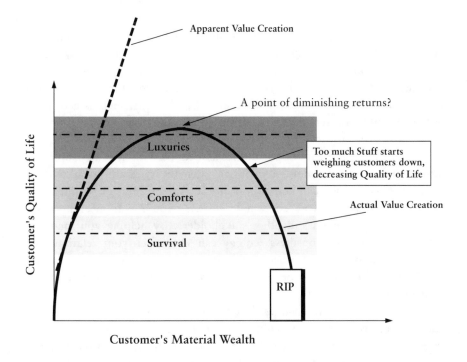

Figure 7.1   Creating Value to Maximize the Customer's Quality of Life
Source: Modified and adapted from *Your Money or Your Life*

the planet are struggling on the upward side of the curve, unable to meet even basic needs.

A sustainable economy ensures that every person is able to meet their basic needs and seeks to maximize the number of people able to reach the uppermost part of the curve where the quality of life is highest. A sustainable enterprise does not, for example, exploit customers' potential dependency on material-intensive products with a short life span or on "addictive" shopping to fill unmet needs by unsatisfactorily acquiring too much. Instead, a sustainable enterprise delivers products and services of high quality and high levels of customer satisfaction. With business ecology, products and services meet these objectives efficiently using life-sustaining flows, time, and space while optimizing value among stakeholders.

The "simplicity" or "down-shifting" movement is a major trend that will affect business and society as a whole. In many ways, it is a revolt against the age of consumerism and an important step toward sustainable development. People are consciously trying to simplify their lives by reducing their consumption and income requirements. This helps them to move back up to the pinnacle shown in Figure 7.1, where their lives are less stressful and more enjoyable. Fulfillment of real needs, such as a balance of purposeful work, healthy relationships, and more discretionary time, are what most of us are seeking.

There are significant opportunities associated with helping people live more simply and have more control over their lifestyles and livelihood. In fact, business ecology can help you position your business or organization favorably in the ecological economy by showing you how to meet the genuine needs of customers and other stakeholders. **By helping you focus on actual value creation, business ecology can strengthen your relationships with your customers and other stakeholders by helping you create enduring, high-quality, closed-loop products and services that genuinely satisfy their needs.**

Table 7.1 compares values development within organizations before and after applying business ecology's mindware. The left side of the table reflects a continuum of values development within organizations. Take a moment to consider whether your organization has clearly defined its core values. If not, consider conducting a systemic appraisal of your organization's values system. The Business Ecology Roundtables^{SM} process includes tools, exercises, and mindware for discovering your organization's deeper, core identity and vision. This knowledge of "who you are" and "what you do" as an organization provides insights into building strategic advantage, including how your organization "fits" within its evolving business ecosystem. The Roundtables^{SM} process includes

- Conducting search workshops and focus groups to uncover your organization's social DNA, cultural history, and values-misalignments
- Using systems lenses, such as society, science and technology, economics and financing, policies and institutions, and the environment
- Interviewing stakeholder representatives to determine how your organization's social DNA affects, and is affected by, its metabolism, niche, and habitat
- Identifying opportunities for integrating sustainability principles into your organization's values system

If values are defined, how well does your organization live by them? Are there aspects of your business or organization's core purpose and iden-

Table 7.1   Organizations Before and After Business Ecology: Values

Before	*After*
1. Shared organizational values are not identified.	1. Shared organizational values are clearly identified.
2. Shared values are identified, but are not serious drivers of organizational decisions and actions.	2. Shared values are recognized as the social DNA of organizations, shaping decisions, perspectives, and behavior.
3. Competing value systems exist within an organization, which may or may not be recognized. This often causes confusion and bad feelings, especially with respect to how success is defined. Organization lacks vision and purpose, and development is fragmented and unhealthy.	3. Shared values clearly define what success means for an organization and shape its development in healthy ways.
4. Shared values are identified, are important organizational drivers, but are not measured or evaluated.	4. Shared values are identified, are important organizational drivers, and are regularly measured or evaluated in a systemic fashion (see figure below).

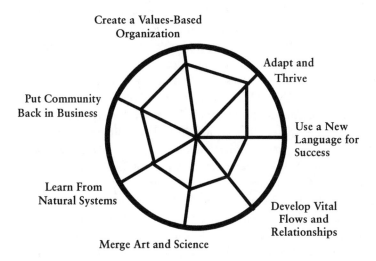

Figure 6.1   Is Your Organization Aligned with Its Values?

© 1998 Business Ecology Associates

tity you would like to change? How can you accomplish such a values shift? Whatever the status of your business or organization, business ecology can help you develop a values-based framework for success. The right side of the table describes the benefits of applying business ecology within your organization.

Figure 6.1, shown in Table 7.1, is a self-referential map of the Business Ecology Network (BEN)'s values that was introduced as an example in Chapter 6. Collectively, the polygon is a composite picture of how well our organization is aligned with its core values. Points on each of the radii measure alignment for each value, with degree of alignment increasing toward the circumference. A combination of quantitative and qualitative indicators and techniques are used to evaluate progress toward our system of shared values, including surveys of members, clients, employees, and suppliers, and detailed assessments of organizational processes and the work environment.

Why is such a tool important to your business or organization? Once established, this self-referential tool can help your organization create a vision, make strategic choices, weigh priorities, balance its system of shared values, flows, and relationships, and align its cultural identity with its short- and long-term planning. Ideally, a business or organization will look for ways to integrate sustainability principles into its values system. In the case of BEN, our core purpose—"To be a catalyst for life-sustaining enterprise"—is grounded in sustainability principles. BEN exists to help organizations transition into sustainable enterprises.

Figure 7.2 compares stakeholder relations before and after applying business ecology within an organization. This systemic tool, also introduced in Chapter 6, can assess your company's multiple stakeholder relations simultaneously. In this case, shareholders, employees, customers, suppliers, communities, the general public, and the environment are plotted to represent the stakeholder community. The points on each of the radii inside the circle—"stakeholder spokes"—are an overall measure of the strength of each stakeholder relationship. This figure is a systemic profile of an organization's relationships with its stakeholder community. That is, points closest to the circumference are the strongest, while the points closer to the center indicate areas that are weak and underdeveloped. What is essential is finding a balance while still achieving important goals. **More and more, successful companies are recognizing that these relationships are interdependent.** Short-term deficiencies or neglect of one relationship, such as community, will have long-term impact on another, such as loyal employees.

BEFORE   AFTER

Figure 7.2   Organizations Before and After Business Ecology: Stakeholder
Relations   © 1998 Business Ecology Associates

In this comparison, the systemic profile on the left represents an organization that excels with shareholders and customers, but needs to strengthen relations with the environment, general public, communities, suppliers, and employees. The one on the right represents an organization with stronger, more balanced relationships with its stakeholder community. Recall that Malden Mills, discussed in earlier chapters, is such an organization. Malden Mills' innovations, quality products and services, sustained earning streams, and resiliency are linked to its healthy stakeholder relations. The right profile, representing a Malden Mills-like organization, is a larger polygon that resembles a circle, indicating greater overall value creation that is more evenly, equitably distributed among stakeholders. (Chapter 6 includes a more detailed discussion of this tool, including how plots on the spokes might be determined.)

Figure 7.3 compares two organizations using two measures of success: one, overall viability, and, two, financial viability. The profiles labeled "overall," like Figure 4.1 in Chapter 4, help you simultaneously assess your organization's life-sustaining flows. Products and services; water, food, materials, energy, and air; people and other organisms; information and ideas; and financial flows are plotted to give a composite picture of overall organizational viability. The nine plots are equally spaced to convey an ideal balance as found in nature. The points on the spokes represent the viability of each of these flows. That is, points plotted closer to the circumference represent flows that are more viable, while those points closer to the center indicate flows that are not as viable or healthy. Similarly, the systemic pro-

BEFORE   AFTER

Financial Viability   *Financial Viability*

Overall Viability   *Overall Viability*

Figure 7.3   Organizations Before and After Business Ecology: Financial and Overall Viability   © 1998 Business Ecology Associates

files labeled "financial" are composite pictures of traditional financial indicators. In this case, seven indicators are plotted: profits, cash flow, market share, earnings/share, management skills, return on assets, return on investment.

Figure 7.3 shows how business ecology can improve your organization's bottom line and overall viability by expanding your perspective from cash flow to life flows. Notice that both the financial and overall viability profiles shown on the right are larger polygons than those to the left, indicating an increase in viability. The reason for the difference is simple: **business ecology's mindware reveals opportunities for ecological efficiency** often missed by bottom-line-driven organizations. For instance, Minnesota-based 3M, the most visionary of the companies studied by Collins and Porras in *Built to Last,* has saved millions of dollars by improving environmental performance through its Pollution Prevention Pays program. Dupont, Monsanto, and others have achieved similar results by preventing pollution and reducing the "ecological footprint" of their products and services.

Business ecology recognizes that each organization may see success differently based on their needs, values, and experiences. Business ecology can be tailored to these specific needs, and incorporates practical, proven business strategies and leading-edge thinking. **Business ecology can save your organization time and money: It is a community of knowledge and experience that synthesizes insights, vision, and professional expertise, including the best thinking from successful business and management leaders.**

Business ecology is the language of sustainable enterprise. It recognizes that our ultimate foundation, the means for supporting our civilization and improving our quality of life, is the Earth. It also recognizes the importance of learning and adopting enduring values, or social DNA, from successful human cultures. Business ecology weaves together the organizing elegance of natural systems and the best wisdom of human cultures to help your organization reach more enduring levels of success.

 4. DEVELOP VITAL FLOWS AND RELATIONSHIPS

Business ecology, as an organic lens and framework, redefines organizational management, accounting, and economic development. It helps you see and develop the vital flows and relationships needed to sustain your business or organization within its environment. For instance, the business ecology lens widens your organization's perspective from "cash flow" to "life flows" and from "accountability to shareholders" to "accountability to stakeholders." Cyclical flows—such as money, energy, materials, information and ideas, products and services, people and other organisms, air, food, and water—sustain economic activity at multiple levels; these include products and services, individuals, processes, firms, communities, and economies. Business ecology is relationship oriented. Your organization can build constructive, closed-loop relationships with other organizations, such as resource exchange networks in a business ecosystem, while enhancing its own ecological efficiency, sense of community, and profitability. Business ecology builds synergistic relationships that can reconnect your busi-

*ness or organization with its community and local, regional, and global
economies and environments.*

Business ecology's mindware, its values-based models for organizational management and design, help you develop vital flows and relationships. How does your business or organization "fit" within developing business ecosystems? What is the best system of relationships and collaborative networks to put in place? Systemic profiles of values, cyclical flows, and stakeholder relations, as discussed in preceding sections, provide a holistic picture of your organization within its environment, including the viability of its vital flows and relationships. In this sense, business ecology is a framework for developing short- and long-range scenario plans for individual companies and business ecosystems. These plans incorporate a new—systemic—way of seeing from different perspectives. What are some of these system lenses? They include society, economics and financing, policies and institutions, science and technology, and the environment. Insights gained from each of these areas are used to develop scenarios that address critical uncertainties, key trends, and predetermined elements (i.e., trends or conditions, such as the U.S. Federal deficit, that are persistent or inevitable because of their momentum).

As discussed throughout this book, systems thinking is crucial to understanding and developing sustainable human systems, including organizations, technologies, and economic systems. With this systems perspective, all flows and relationships take on a new relevance and value. As discussed in Chapter 4, cyclical, life-sustaining flows contribute to your organization's value-creation process. These flows include: products and services; water, food, materials, energy, and air; people and other organisms; information and ideas; and money. Your organization's viability and ecological efficiency can be improved by conducting ecological assessments and applying strategies, such as water and energy conservation, supporting creative thinking, and closing the loop with materials, products/services, and by-products.

Strengthening stakeholder relations builds your organization's viability. Business ecology recognizes that value creation extends beyond the customer and shareholder to a broader stakeholder community. For example: customers experience recognizable benefits from products and services that actually meet or exceed their needs, desires, and expectations; shareholders and creditors receive returns on investments; employees are compensated with salaries, health benefits, and equity shares; federal, state, and local governments receive tax revenues both from the company and its employees; and nonprofit groups receive tax-deductible donations. These value flows, in turn, support value creation in other organizations.

Business ecology, as a values-based organizing framework, recognizes that core purpose and values, or social DNA, shape how an organization develops vital flows and relationships within its environment. These organizational flows and relationships, like natural systems, are fractal; they exist at different levels of organization, such as firm, community, and regional economy. **Shared values strengthen the organizing capacity of sustainable enterprises and business ecosystems by creating a strong social fabric for collaboration.**

Information technologies—such as remote sensing, "smart" matchmaking software; geographic information systems; and Internet technologies—will have a crucial role in "growing" sustainable businesses and business ecosystems. The Internet, for instance, is helping many businesses and grassroots organizations share a sense of community as well as information on proven strategies and technologies, ways of collaborating on projects, cutting edge research, and even how to avoid costly travel. Another example is "smart" match-making software, where electronic "brokers" make valuable links among suppliers and users of information, by-products, and other resources. In the context of business ecology, such software can help track, exchange, and manage energy, materials, water, products and services, and other vital flows among participants in a business ecosystem. **Information technologies, acting much like the nervous, sensory, and endocrine systems in our bodies, will help regulate cyclical, value-creating flows within the ecological economy.**

Business ecology helps you solve problems and develop new opportunities. It strengthens your company's viability by helping you:

(1) see and improve your company's metabolism;
(2) "close the loop" with other companies;
(3) develop new products and services from residual flows; and
(4) create multiple profit cycles.

For business ecosystems, such as those profiled in Chapter 5, to thrive and prosper, they or their organizing agents must understand the ecological flows of individual companies and their neighbors. When such interconnections are made, everybody wins. This leads to profitability, healthy stakeholder relations, and a cleaner environment. Combined with the knowledge and insight to know what to do, and the creativity and boldness to make it happen, business ecology can make closed-loop resource recovery work and reuse a lucrative opportunity for your company.

 ## 5. MERGE ART AND SCIENCE

Business ecology is both art and science. It incorporates the ability to see patterns, systemic relationships, different perspectives, and express creative imagination, which are traditionally the skills of an artist. It is also a science, in that it draws conclusions and creates models for organizations based on a close observation of natural systems. It integrates leading-edge thinking and proven success strategies into mindware—organic management models that save you time, money, and resources. As a synthesis, business ecology distills the best thinking from diverse fields and insights from natural systems, making these innovations more understandable, applicable, and useful to leaders and managers such as yourself.

Business ecology merges art and science to help you see from different perspectives and apply these insights to solve problems and create opportunities. Observing patterns, systemic relationships, varying perspectives, and expressing ideas and creative imagination are visual artistic skills you can incorporate through business ecology into your day to day and long-term organizational management and design. Coupled with an inquiring, problem-solving scientific perspective business ecology builds measurable strategies, models, and mindware for organizational management that emulate natural systems.

Business ecology's approach to organizational management and design is fractal. Just as shapes and systems in nature, such as the spirals of seashells and galaxies, repeat themselves on all scales, so too do an organization's shapes and systems. In this case, the underlying design principles for the spiral shape are the same no matter what the scale, just as the underlying principles—core values—for an organization's design remain constant. What is useful about this perspective when managing your business or organization, is **that it allows you to see that relationships, opportunities, and problems are related—rather than fragmented. A solution that works in one area may in fact be a fit in all, once you recognize that it is only a matter of scale.** Business ecology is based on the belief that organizational management, like scientific research or artistic expression, requires not only factual research and a cohesive design, but a well-documented leap of faith—the inspiration of the creative imagination—to discover solutions.

6. LEARN FROM NATURAL SYSTEMS

Business ecology is based on the organizing elegance of natural systems, the success secrets that have accumulated over 3.5 billion years of evolution. Business ecology, by emulating natural systems design, offers several answers: articulating your organization's core genetic code, i.e., core purpose and values, life-cycle thinking, resource exchange relationships, and models for organizational design that encourage innovation, resiliency, and adaptability. These models are fractal, they apply to all sizes and scales of business and organization and are relevant for both day-to-day management as well as long-term strategic planning.

How can you tap into the organizing elegance of natural systems? Spend more time enjoying the natural environment. Observe systemic balance, patterns, relationships, and design characteristics. Successful organisms—and organizations—reflect robust ecological design and adaptive instincts. Qualities that contribute to adaptive success include: numerous feedback loops to sense and interact with the environment; strong symbiotic relationships with other organisms, ability to withstand stress and return to a systemically balanced state; and a diverse gene pool to draw upon in rapidly changing conditions.

The Natural Step (TNS), founded by the Swedish physician Dr. Karl-Hendrik Robért, weaves insights from nature and common sense techniques for building consensus to focus attention on sustainable development. It is a global movement to help people and organizations see and respond to the sustainability challenge. Robért was concerned that increasing numbers of cancer cases among children were tied to environmental conditions. In his frustration with scientific disagreement about the underlying causes and possible solutions, he founded a consensus-building process among scientists, business people, engineers, politicians, artists, doctors, architects, and economists to determine the underlying principles of sustainability. The King of Sweden and Swedish business and entertainment leaders have strongly supported TNS.

Business leader Paul Hawken, author of *Ecology of Commerce*, recognized the ingenuity and political savvy of this approach and is working with Robért and others to help propagate TNS principles in North America, Europe, and around the world. Already, companies such as Monsanto, IKEA, McDonald's, and Interface, Inc., have undergone the training and are strategically realigning their businesses toward sustainable enterprise. Below are the four system conditions for sustainability identified by TNS:

System Conditions for Sustainability

1. Substance from the Earth's crust must not systematically increase in the biosphere. This means that fossil fuels, metals, and other minerals must not be extracted at a faster rate than their redeposit and reintegration into the Earth's crust.

2. Substances produced by society must not systematically increase in nature. This means that substances must not be produced faster than they can be broken down and reintegrated into natural cycles.

3. The physical basis for the productivity and diversity of nature must not be systematically deteriorated. This means that the productive surfaces of nature must not be diminished in quality or quantity, and that we must not harvest more from nature than can be recreated and renewed.

4. There needs to be fair and efficient use of resources with respect to meeting human needs. This means that basic human needs must be met with the most resource efficient methods possible, including equitable resource distribution.

Guided by these principles, TNS organizations are challenged to create their own strategies for sustainable enterprise, based on their own creativity, knowledge, and experience. Business ecology integrates TNS principles into its models and mindware, and brings a specific focus on sustainability: organizational management and design.

 ### 7. PUT COMMUNITY BACK IN BUSINESS

Business ecology puts "community" back in business. Community lies at the heart of a healthy, life-sustaining business or organization, and is the essence of our spiritual well-being. Business ecology redefines success and is a powerful agent for restoring community within companies and organizations and the communities of which they are a part. Its mindware for sustainable enterprise offers models based on natural systems that incorporate the need for community among successful, surviving organisms— and organizations. Community is all about connections—to ourselves and each other.

Successful organizations adhere to a simple but vitally important principle: **build community in your organization by identifying and committing to core values, and the profits will follow.** Business ecology recognizes this principle, but goes a step further in asking: Are your core values consistent with sustainability values? By drawing on the organizing elegance of natural systems and the best wisdom of human cultures, business ecology helps you build community within your business or organization. Here are sixteen sustainability principles, developed by Four Worlds International (based in British Columbia), that provide a solid foundation for building community at all scales:

Four Worlds Principles for a Sustainable Society

These sixteen principles for building a sustainable world emerged from a twelve-year process of reflection, consultation, and action within tribal communities across North America. They are rooted in the concerns of hundreds of aboriginal elders and leaders of thought, as well as in the best thinking of many nonaboriginal scholars, researchers, and human and community development practitioners.

These principles are part of business ecology's mindware for the process of healing and developing ourselves (mentally, emotionally, physically, and spiritually), our human relationships (personal, social, political, economic, and cultural) and our relationship with the Earth. They describe the way we must work and what we must protect and cherish.

> We (Four Worlds International) offer these principles as a gift to all who seek to build a sustainable world.

THE GUIDING PRINCIPLES

Starting from within, working in a circle, in a sacred manner, we heal ourselves, our relationships, and our world.

STARTING FROM WITHIN

Development Comes from Within
The process of healing and development unfolds from within each person, relationship, family, community, or nation.

Vision
A vision of who we can become is like a magnet drawing us to our potential. Where there is no vision, there can be no development.

Culturally Based
Healing and development must be rooted in the wisdom, knowledge, and living processes of our cultures.

Interconnectedness

Because everything is connected to everything else, any aspect of our healing and development is related to all the others (personal, social, cultural, political, economic, etc.). When we work on any part, the whole circle is affected.

WORKING IN A CIRCLE

Growth and Healing for the Individual, the Family, and the Community Must Go Hand in Hand

Working at one level without attending to the other is not enough. Personal and social development as well as top-down and bottom-up approaches must be balanced.

Unity

We need the love, support, and caring of others to heal and develop ourselves. Unity is the starting point for development, and as development unfolds, unity deepens.

Participation

People have to be actively engaged in the process of their own healing and development. Without participation, there can be no development.

Justice

Every person must be treated with respect as a human being and a child of the Creator, regardless of gender, race, culture, religion, or any other reason. Everyone should be accorded the opportunity to fully participate in the processes of healing and development, and to receive a share of the benefits.

IN A SACRED MANNER

Spirituality

Spirituality is at the center of healing and development. Connection with the Creator brings life, unity, love, and purpose to the process, and is expressed through a heart-centered approach to all that we do.

Harmonizing with Natural Law

Growth is a process of uncovering who we truly are as human beings in harmony with the natural laws of the Universe.

Walking in Balance

Codes of morality, ethics, and protocol teach us how to walk the road of life in a good way. Violating moral and ethical boundaries can destroy the process of healing and development.

Working from Principle

Our plans and actions are founded on our deepest understanding of the principles that describe how the universe is ordered and how healing and development unfold.

WE HEAL OURSELVES, OUR RELATIONSHIPS, AND OUR WORLD

Learning

Learning to live in ways that promote life and health is the essence of our development. Our primary strategy is therefore the promotion of this type of learning.

Sustainability

When we take actions to improve our lives or the lives of others, it is critical to avoid undermining the natural systems upon which all life depends and to work in ways that enhance the capacity of people to continue in the path of their own healing and development.

Move to the Positive

Solving the critical problems in our lives and communities is best approached by visualizing and moving into the positive alternative that we wish to create, and building on the strength we already have, rather than giving away our energy fighting the negative.

Be the Change You Want to See

In all of our actions, we seek to be living examples of the changes we wish to see in the world. By walking the path, we make the path visible.

These principles provide steps for transforming ourselves and our organizations to create a sustainable future. The challenges before us are colossal, but so are the opportunities. Dee Hock, founder and CEO Emeritus of Visa International, challenges all of us with this vivid, hopeful vision for our future:

> We are at that very point in time when a 400-year-old age is dying and another is struggling to be born—a shifting of culture, science, society, and institutions enormously greater than the world has ever experienced. Ahead, the possibility of the regeneration of individuality, liberty, community, and ethics such as the world has never known, and a harmony with nature, with one another, and with the divine intelligence such as the world has never dreamed.

Business ecology meets the organizational challenges posed by this transformation, the inevitable transition to a sustainable, ecological economy. It is a community of knowledge and experience that synthesizes insights, vision, and professional expertise, including the best thinking and wisdom from diverse cultures, fields, and successful business and management leaders. Business ecology is, in fact, a new way of doing business, it is mindware for the new millennium. **By modeling organizational management and design on natural systems and the values of sustainability, business ecology can help your business or organization transition to the future.** It is a comprehensive, values-based organizing framework that helps you link profitability, stakeholder relations, life-cycle thinking, and environmental performance. Business ecology gives you an ecological lens to see vital, systemic relationships and life-sustaining flows within your business or organization and its environment. It also shows you how to see your business or organization's design and purpose from different perspectives and apply these insights to solve problems and create opportunities. It is our conviction that by using business ecology, your business or organization will be strategically positioned in the new, emerging ecological economy. This is because business ecology puts "community" back in business, helping us to reconnect to basic truths, each other, and the web of life.

Appendix

ORGANIZATIONAL MODELS PRECEDING THE BUSINESS ECOLOGY MODEL

The "Flush Through" Model

Figure A.1a represents the business climate in the United States before 1970, a date which marks the formation of the U.S. Environmental Protection Agency. The EPA was created by the U.S. Congress to help remedy serious pollution problems caused by irresponsible dumping of wastes. You may recall several of the pressing environmental problems of the late 1960s. In 1969, for instance, one could actually smell Lake Erie from Niagara Falls, New York, to Detroit, Michigan. Anyone could tell there was a serious problem with the Lake; a person did not need to be a scientist. The late 1960s was also a time when once-pristine rivers were catching on fire. Proximity to the Cuyahoga River in Cleveland, for example, did not lower insurance costs to businesses, it increased them. At its worst, proximity to the Cuyahoga became a reason for insurers not to insure some businesses at all.

The reasons for these and other disasters were simple. The business climate before 1970 encouraged simply transforming material and energy into commercially viable products or services, while paying little or no attention to "nonproducts" discharged into the environment. Virgin resources were simply "flushed through" the production system in a single pass. Anything that was not product, was simply "nonproduct" or waste.

Figure A.1a is a simplified model of business operations in the U.S. before 1970. Note that nonproduct residues are shown as discharges or emissions to the environment, originating in the production box. In the "Flush Through" model, several industries do what they have always done: take resources from the environment, use what is needed, and throw back anything that is not needed. In this system, commodities that are commonly available, such as clean air and clean water, are devalued to zero, partly because they are so abundant. And since this commodity-costing system

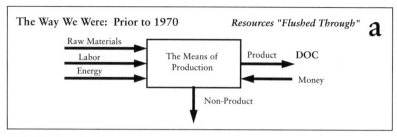

The Way We Were: Prior to 1970 *Resources "Flushed Through"* **a**

Raw Materials
Labor
Energy
→ The Means of Production → Product → DOC
← Money
↓ Non-Product

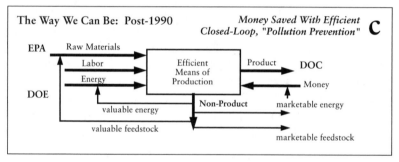

The Way We Were: 1970-1990 *Responding with "Polluter Pays" Costly Laws & Regulations* **b**

EPA
Raw Materials
Labor
Energy
DOE
→ The Means of Production → Product → DOC
← Money
↓ Multi-Media Waste

1970 NEPA and EPA formed

Air   Land   Water

Clean Air Act   RCRA, TSCA and others   Clean Water Act, etc.

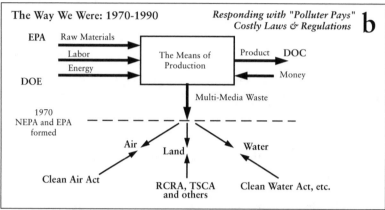

The Way We Can Be: Post-1990 *Money Saved With Efficient Closed-Loop, "Pollution Prevention"* **c**

EPA
Raw Materials
Labor
Energy
DOE
→ Efficient Means of Production → Product → DOC
← Money
← marketable energy
Non-Product
valuable energy
valuable feedstock
marketable feedstock

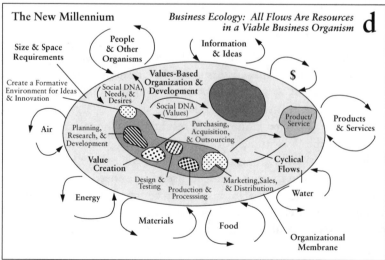

The New Millennium *Business Ecology: All Flows Are Resources in a Viable Business Organism* **d**

Size & Space Requirements
People & Other Organisms
Information & Ideas
$
Create a Formative Environment for Ideas & Innovation
Values-Based Organization & Development
Social DNA, Needs, & Desires
Social DNA (Values)
Product/ Service
Products & Services
Air
Planning, Research, & Development
Purchasing, Acquisition, & Outsourcing
Value Creation
Cyclical Flows
Water
Design & Testing
Production & Processsing
Marketing, Sales, & Distribution
Energy
Materials
Food
Organizational Membrane

© 1998 Business Ecology Associates

placed no value on these resources, it was implied that they should be available to all for free, perhaps even taken for granted. By extension, this thinking created a sense of entitlement, the belief by many industries in a "right" to pollute.

However, the private appropriation of public resources created public inequity. Consumers paid once for the value that manufacturers added to natural resources in the form of manufactured goods, and a second time for restoring public health from the effects of pollution in the form of increased medical costs. On top of these costs of increased human sickness, consumers paid a third time for human welfare costs, paint damage, and dissolving national historic monuments, for example, and yet a fourth time for environmental and aesthetic losses, such as loss of visibility, less clarity in the atmosphere, damage to vistas and plant and animal life through loss of habitat. Some argue that a fifth cost stems from the damage done to the spiritual relationship between Man and Earth.

The "Polluter Pays" Model

When the health and environmental prices became high enough, public outrage translated into votes, and then into proenvironmental policies and statutes. This reaction to the degradation of public resources by private industry resulted in a second and distinct phase in the domestic business climate in the 1970s. It ushered in the age of "Polluter Pays," an approach to environmental protection made possible—in fact inevitable—by the behavior of private industry. Some industry leaders allowed themselves to focus exclusively on "cash flow" as a surrogate indicator of the health of all other corporate flows. Some objected strongly to environmental regulation, believing that "cash is king"—and that nothing else mattered. Not having to live in the communities with the problems that they had created, these industry leaders viewed environmental statutes as a transfer of wealth away from their companies. In fact, the environmental statutes of the 1970s and 1980s did seek to stop a wealth transfer that was already occurring: from the public wealth of clean natural resources to the private gain of polluting companies. A key goal of the new statutes was to minimize negative effects on human health and promote the concept of the "polluter pays," which was viewed by the public and by Congress as equitable. "Pol-

Figure A.1 (Opposite) **Four Organizational Models: Changing Relationships Among Business, Society, and Environment**

luter pays" made sense to everyone except the polluters. From their perspective, the statutes caused wealth to be transferred away from private profits and toward the general public to pay for a social "good." Some viewed this as "un-American" because it hampered free enterprise. The loudest objections came from those who still operated in the previous "Flush Through" model. In "Flush Through," public resources of air and water were free, and the polluter felt entitled to use them—and to pollute them when necessary. The cost of degraded natural resources was totally external to the costs of production.

During the 1970s and 1980s, public policy reflected the fact that people recognized that their own personal health was inextricably linked to the quality of the air that they breathed and of the water that they drank. In 1970, a strong amendment to the Clean Air Act reaffirmed Congress's intent to "protect human health and welfare." A Clean Water Act asserted a goal that U.S. waters should be "fishable and swimmable." Deadlines for compliance with the new ambient air standards were set, and then missed. As an example, the amended Clean Air Act sought compliance with the new National Ambient Air Quality Standards by 1975. By 1977, it was clear that the existing statute needed to be strengthened in order to fulfill the intent of Congress.

But why did the nation require the 1977 amendments to the Clean Air Act? Several reasons are apparent. One is that the institutional infrastructure needed for the clean-up job had to be created largely from scratch. While parts of the U.S. Public Health Service and other existing government organizations were drawn into the EPA, the national scope of the mission was large. Regulatory infrastructure and case law were lacking. Also, the problems that the country faced were far greater than many imagined. Few understood the issues surrounding implementation of the new statues. Few understood the depth of industry's resistance. Few would imagine the ensuing tugs-of-war over State's Rights, the near collision with the Appointments Clause of the U.S. Constitution, the endless court suits where experts represented opposing sides with equal conviction; the seemingly endless contention over findings of culpability, and most importantly, who would pay for the clean-up. The resistance of industry to solve problems that they themselves created underscores the fact that values do indeed shape economic systems. In this case, industry valued profit and the cheap "flushing through" of resources.

Figure A.1b shows how public pressure and the force of legislation reclassified the "nonproduct" discharge of the earlier paradigm and called it "pollution." The government's response was then to further divide the pollution by type, by separating the pollutants into a taxonomy of "media" such as land, air, and water. Through this new "media" lens various statutes were fashioned to resolve media-specific environmental problems. So when the Clean Air Act Amendments of 1970 pushed back on discharges into the air, two significant things happened. The first was immediate confrontation between businesses and the new authorities; and the second was a set of unintended consequences. In controlling one pollutant in one media, another problem might occur in another media. Sometimes an air pollutant might be converted to a water pollutant or to a land pollutant. For example, catching particulate matter from smokestacks resulted in new disposal problems on land for the newly captured fly-ash.

National and state institutions responded by creating more legislation, moving from media to media, and from problem to problem. Neither industry nor government was yet looking at the problem comprehensively. Unfortunately, perhaps inevitably, industry and government viewed themselves as opponents, and attempts to clean up were met with attempts to evade. A good example of this can be found in the 1970s with air pollution. Since the National Ambient Air Quality Standards were expressed as a concentration—so much pollution per volume of air—industries essentially played games with the system instead of cleaning up. Cars were equipped with air pumps to simply mix a lot of cleaner air with the tailpipe gases to lower the concentration of pollutants. Power plants were refitted with tall stacks to better use the diffusive capability of the atmosphere. Emissions were not reduced so much as mixed with a greater volume of clean air so that "compliance" could be achieved. In fact, there was little profit motive in changing the current way of thinking. The "Flush Through" paradigm was so deeply entrenched and so lucrative for industry that it dominated attempts to comply with the new regulations.

The "Pollution Prevention" Model

By 1990, a new "Pollution Prevention" model evolved within state and national institutions and industry. Rather than forcing compliance to regulations, the pollution prevention approach looked more holistically at the means of production—and took into account what is of value to all parties, especially the need for industry to minimize costs and the need for citizens not to have their environment violated. As is the usual case with shifts in thinking, there is no line separating one—control by regulations—from the other—pollution prevention. In fact, both the "Pollution Prevention" paradigm and earlier ways of addressing environmental problems operate simultaneously. Those companies that have taken a more holistic view, considering all the flows in and around their organization, are, in fact, becoming more profitable. They are creating a new force in the marketplace that spells only one thing for competitors: change or vanish. This shift in thinking underscores why the time is ripe for applying the principles of business ecology.

How are these companies succeeding? Figure A.1c shows a return to "Flush Through" in one significant respect. The "Pollution Prevention" model also views discharges as "nonproduct." By thinking of pollution in this way, it is easier to see that pollution stems simply from inefficiency in converting material and energy into commercially viable products or services. The greater the pollutant discharge, the greater the inefficiency of conversion. What is essential is that, in the business ecology scenario (Figure A.1d), government's role changes dramatically from "enforcer" to "catalyst," and industry changes from "polluter" to "eco-efficient enterprise."

A COMPARISON OF HYPOTHETICAL NICHE ENTERPRISES

Here are some hypothetical examples, based on actual facts and conditions, that illustrate "niche" in the context of business ecology. The three examples—a franchise bagel sandwich shop, a community-based grocery store, and an ecologically efficient bakery/brewpub/restaurant—show how, in the same habitat, different niches are filled by similar and sometimes competing businesses. The bagel bakery/sandwich shop shows efficient organizational design based on franchise/owner social DNA; targeting customers based on income, schedule and proximity; a dependency on commuters and employees of the medical and professional fields; a national/international supply web; and a flow of automobile and pedestrian traffic. The grocery store/bakery shows a diverse enterprise built on the values and core purpose, i.e., the social DNA, of the owners, community, and stakeholders in the neighborhood it serves. It also illustrates development based on community needs and desires; a regional/local supply web; responsiveness to neighborhood flows and activities, such as convenient pedestrian access; a diverse selection of goods and services; and delivery. Finally, the bakery attached to the brewpub/restaurant depicts an enterprise that's design stems from the social DNA of an anchor business and its neighborhood; efficient use of space, food, energy, and employees; regional/local closed-loop flows; and convenient automobile and pedestrian access.

Location: Inner West Street, Annapolis, Maryland, U.S.A.

Inner West Street, a seven-block area along the original land and train route into Annapolis, Maryland, is undergoing economic revitalization. This area of town is on the edge of the fashionable historic and waterfront areas of downtown Annapolis; it is approximately one mile from an outer perimeter of highways and development. It is a mixture of newly redeveloped commercial fronts that cater to mid- to upper-income customers. These include: a southwestern restaurant, a coffee and wine bar, a fast-food restaurant, a brewpub, an art gallery, and a florist. There are also abandoned businesses—a movie theater, pharmacy, seafood store, and gas station are now remnants of a business district that once met the needs of its local community. Rent is somewhat cheaper than other sections of town, but there are also concerns about its uneven appearance and potential crime. The site is very close to a low-income minority neighborhood that is struggling to revive itself. The Clay Street neighborhood has high unemployment and is underserved in basic areas such as staple groceries and basic, affordable medical care. Many residents cannot even afford cars. Some funding and other assistance is available from local and state groups, private foundations, and the federal government for business start-ups. Customers to Inner West St., who may come from outside the community, travel to the establishments by car and park on the streets or in the local, subsidized parking garage. The following scenarios represent distinctly different niches that could develop here for food-related businesses.

Scenario A: Bagel Bakery/Sandwich Shop

In scenario A, an entrepreneur decides to open a carry-out and eat-in bagel bakery and sandwich shop, which caters to customers employed at the recently expanded medical center and renovated professional offices along West Street and to tourists who visit Annapolis. Her market survey indicates that many of these customers are both highly paid and in a hurry. Many work long hours, including weekends and holidays. She decides to open a bagel bakery/sandwich shop that uses low-fat, health-conscious ingredients, condiments, and equipment supplied by a national franchise, Bagels Etc. It also features desserts, coffee, espresso, and herbal teas. A small counter with stools is available for eating near the entrance. Her market study, however, indicates that most customers will want to order their food to go. The store is open from 5:30 a.m. to 3:30 p.m. Monday through Friday; 8:00 a.m. to 3:30 p.m. Saturday; and is closed on Sunday. She hires employees from local high schools and colleges through ads in the local paper. The business prospers.

Scenario B: Community Grocery Store/Bakery

In scenario B, some residents from the Clay Street neighborhood decide to open a community grocery store/bakery, called ShopHere, to serve residents in Inner West Street and adjoining areas. The founders take full advantage of financial and business planning assistance provided by the federal and local governments. ShopHere features local produce, meats, and fish from Maryland and the mid-Atlantic region, including vegetables, fruits, herbs, and flowers from the local urban garden project in the Clay Street neighborhood. The bakery caters local church functions and parties at professional offices. It is the only grocery store within the city limits that residents can walk to. Future plans call for adding services such as a pharmacy and video rentals. The store is open Monday through Saturday from 9:00 a.m. to 10:00 p.m. and Sunday from 10:00 a.m. to 5:00 p.m. The business becomes a significant employer, a social gathering place, the training ground for future entrepreneurs, and a source of neighborhood pride. Profits from the business help establish an education fund for children in the neighborhood.

Scenario C: Bakery Associated with BrewPub/Restaurant

Scenario C involves the owner of Brady's brewpub/restaurant, who sees an opportunity related to his current business. He reads about the Seattle Spent-Grain Baking Company, a business that collects spent grain—a by-product from the brewing process—from local microbreweries to use in the baking of fresh breads. The breads are nutritious, tasty, and distinctive in color and texture. Excess grains from the Spent-Grain Baking Company are sold to local farmers. Brady has additional space available and would like to provide his own fresh baked goods to his customers. He opens a bakery adjacent to his current establishment and finds that excess energy from baking and brewing can be exchanged to lower operating costs. His bakery features breads, rolls, breadsticks, and pretzels that use spent grain from his brewpub. It serves both the restaurant/brewpub and customers from local businesses and the neighborhood. He trains employees from his brewpub/restaurant to operate the bakery and uses underutilized ovens from the restaurant kitchen during off hours, such as early morning. He sells excess spent grains to local farmers for livestock feed. In turn, he purchases cheeses and organic hops from the farmers. Cheese from the farm is sold directly and used by the restaurant for making cheese breads, salad toppings, and sandwiches. The hops are used in the brewing process. Brady himself staffs the carry-out bakery, which he opens earlier than the restaurant/brewpub to help cover the costs of rental space and to reach morning customers. Store hours are from 7:30 a.m. to 8 p.m. Tuesday through Friday; 7:30 a.m. to 5 p.m. Saturdays; 7:30 a.m. to 4 p.m. Sundays; and closed Mondays.

 ## THE BUSINESS ECOLOGY NETWORK

The Business Ecology Network (BEN), founded in 1995, is a catalyst for life-sustaining enterprise. BEN is a learning community for leaders and managers who want to apply a new way of thinking—business ecology—to create new, sustainable opportunities for their businesses and nonprofit organizations.

A New Field

Business ecology is a new field that synthesizes centuries of cultural wisdom, a close observation of natural systems, the values of sustainability, and proven success strategies, such as strategic planning and total quality management. It also draws on such leading-edge thinking as organizational learning, industrial ecology, and ecological economics. The business ecology model is mindware for all types of businesses and organizations. It emulates natural systems design and provides an elegant, relationship-oriented approach that reveals how your organization really works.

Vital Flows and Relationships

Business ecology measures not just financial but overall viability by revealing vital flows and relationships that sustain your business or organization. It is a lens for seeing those intangible elements of your business or organization's design—such as stakeholder relations, core values, community, value-creation cycles, and innovative thinking—that are essential factors shaping its success. Business ecology helps you integrate profitability, values-based management, stakeholder relations, life-cycle thinking, and environmental performance—giving your organization a natural edge in the emerging ecological economy.

BEN offers two programs:

Business Ecology Roundtables^{SM} An innovative series of mindware, tools, and workshops on organizational management and design tailored to fit individual and business ecosystem needs. The Roundtables are designed for businesses and organizations that want to transition towards a sustainable future. You can observe natural systems, identify and align your organization with its core values, build strong relationships, improve your organization's ecological efficiency, develop closed-loop behavior and multiple value-creation cycles, and merge profitability and stakeholder relations with a commitment to the local environment and community as well as a larger, global web of resources and relationships.

Business Ecology News^{SM} includes a quarterly newsletter, *Main Street Journal*, and an Internet Web site—**http://naturaledge.org.**

Contact BEN today to receive a copy of the *Main Street Journal*, more information on the Roundtables, and find out how to become a catalyst for life-sustaining enterprise:

Business Ecology Network
P.O. Box 29
Shady Side, Maryland 20764-9546
Tel.: 410-867-3596
Fax: 410-867-7956

Joe Abe, President
Trish Dempsey, Communications Director
Gregg Freeman, Development Director

Bibliography

Abe, Joseph M. "Clean Technologies and Ecoindustrial Parks," *In Business,* November/December 1995.

Allenby, Braden. "White Paper on Sustainable Development and Industrial Ecology." Unpublished report by Institute of Electrical and Electronics Engineers, 1995.

Alperovitz, Gar, and Jeff Faux. *Rebuilding America.* New York: Pantheon Books, 1984.

Annual Energy Outlook 1998. Washington, DC: Energy Information Administration, U.S. Department of Energy, December 1997.

Bassett, David. "Pollution Prevention Strategy for the Energy Sector." Unpublished report prepared for the U.S. Environmental Protection Agency, 1991.

Bassett, David. "The Promise of Prevention." Unpublished report prepared for the Vice President of the United States, 1993.

Benson, David. "Columbia's School for Business Revolutionaries," *Fast Company,* August-September 1996.

Bushnell, G. H. S. *The First Americans.* New York: McGraw-Hill, 1968.

Business and the Environment. Arlington, MA: Cutter Information Corporation, October 1997.

Callenbach, Ernest, Fritjof Capra, Lenore Goldman, Rudiger Lutz, and Sandra Marburg. *Ecomanagement.* San Francisco: Berrett-Koehler, 1993.

Case, John. "The Open Book Revolution," *Inc.,* June 1995.

Clark, Grahame. *The Stone Age Hunters.* New York: McGraw-Hill, 1967.

Clark, Michael. "Malden Mills Weaves Success From Strong Ties to Workers and Community," *Journal of Innovative Management* (Vol. 2, No. 1), 1996.

Cohen, J. E. *How Many People Can the Earth Support?* New York: Norton, 1995.

Collins, James C., and Jerry I. Porras. *Built to Last.* New York: Harper Business, 1994.

Couturié, Bill, dir. *Earth and the American Dream.* 70 min., Home Box Office Films, 1992.

Covey, Steven. *Seven Habits of Highly Effective People.* New York: Simon and Schuster, 1990.

de Geus, Arie. *The Living Company.* Boston, MA: Harvard Business School Press, 1997.

Dominguez, Joe, and Vicki Robin. *Your Money or Your Life.* New York: Penguin, 1992.

Edwards, Betty. *Drawing on the Right Side of the Brain*. Los Angeles: J. P. Tarcher, 1979.

Estes, Ralph. *Tyranny of the Bottom Line*. San Francisco: Berrett-Koehler, 1996.

Finney, Martha, and Deborah Dasch. *Find Your Calling, Love Your Life*. New York: Simon and Schuster, 1998.

Gablik, Suzi. "A Few Beautifully Made Things," *Common Boundary*, 1995.

Gallagher, Winifred. *The Power of Place*. New York: Harper Collins, 1993.

Gamst, Frederick C. *The Qemant: A Pagan-Hebraic Peasantry of Ethiopia*. New York: Holt, 1969.

Gillette, Becky. "Turning Tires Into Furniture," *In Business*, January-February 1997.

Glenn, Jerome C. *Future Mind: Artificial Intelligence*. Washington, D.C.: Acropolis Books, 1989.

Gradel, T. E., and B. R. Allenby. *Industrial Ecology*. Englewood Cliffs, NJ: Prentice Hall, 1995.

Hall, Sue. "Sustainable Partnerships," *In Context*, Summer 1995.

Harmon, Willis. "The Rebirth of Business: Shifting to the New Economy." Conference brochure. San Francisco: World Business Academy, 1996.

Hart, Stuart L. "Strategies for a Sustainable World," *Harvard Business Review*, January-February 1997.

Hawken, Paul. "Ben & Jerry's Social Performance Report," South Burlington, VT: Ben & Jerry's, 1994.

Hawken, Paul. *The Ecology of Commerce*. New York: Harper Business, 1993.

Henderson, Hazel. *Building A Win-Win World*. San Francisco: Berrett-Koehler, 1996.

Henderson, Hazel. *Politics of the Solar Age: Alternatives to Economics*. Garden City, NY: Anchor Press/Doubleday, 1981.

Hertsgaard, Mark. "The Cost of Global Climate Change," *World Business Academy Perspectives*, vol. 10, no. 4, 1996.

Hock, Dee W. *Basic Text/Speech*. Unpublished. 1978.

Jolly, Clifford J., and Fred Plog. *Physical Anthropology and Archeology*. New York: Knopf, 1976.

Journal of Industrial Ecology (Vol. 1). Cambridge, MA: M.I.T. Press Journals, 1997.

Kiuchi, Tachi, and Bill Shireman. *Invitation to Join Futures 500*. http://www.globalff.org, 1997.

Kunstler, James. *The Geography of Nowhere*. New York: Simon & Schuster, 1993.

Levine, Joseph S., and Kenneth R. Miller. *Biology: Discovering Life*. Lexington, MA: D.C. Heath, 1991.

Lulic, Margaret A. *Who We Could Be at Work*. Boston, MA: Butterworth-Heinemann, 1996.

Lindbergh, Anne M. *Gift From the Sea*. New York: Pantheon Books, 1955.

Magretta, Joan. "Growth Through Sustainability: An Interview with Monsanto's CEO, Robert B. Shapiro," *Harvard Business Review*, January-February 1997.

Main Street Journal (vol. 1 & 2). Annapolis, MD: Business Ecology Network, 1997.

Moore, James F. *The Death of Competition.* New York: Harper Business, 1996.

National Air Pollutant Trends, 1900–1994, U.S. Environmental Protection Agency/ OAR/OAQPS, October 1995.

O'Brien, Kathleen. "How an Ecopark Will Work," *In Business,* November/December, 1996.

Papazian, Charlie. *The New Complete Joy of Home Brewing.* New York: Avon Books, 1991.

Parker, Thorton, and Theodore Lettes. "Using Knowledge to Create Value." Unpublished paper. Bethesda, MD: Growth Cycle Design, Inc., 1997.

Pillay, T. V. R. "The Challenges of Sustainable Aquaculture." *World Aquaculture,* vol. 27(2), June 1996.

Product Stewardship Advisor. Arlington, MA: Cutter Information Corporation, May 1997.

Roszak, Theodore, Mary Gomes, and Allen Kramer, ed., *Ecopsychology: Restoring the Earth, Healing the Mind.* San Francisco: Sierra Club Books, 1995.

Sachs, Ignachy, and Dana Silk. *Food and Energy: Strategies for Sustainable Development.* Tokyo: United Nations University Press, 1990.

Senge, Peter. *The Fifth Discipline.* New York: Doubleday, 1990.

Skillicorn, Paul, William Spira, and William Journey. *Duckweed Aquaculture.* Washington, D.C.: The World Bank, 1993.

Steutiville, Robert. "The Power of Purchasing," *In Business,* 1993.

Tibbs, Hardin. *Industrial Ecology: An Environmental Agenda for Industry.* The Global Business Network, 1993.

Waters, Alice. Letter to Bill Clinton, President of the United States, December 9, 1995.

Waters, Alice. Letter to Bill Clinton, President of the United States, December 17, 1996.

Webber, Alan M. "Think You're Smarter Than Your Computer?" *Fast Company,* October/November, 1996.

Webber, Alan M. "XBS Learns to Grow," *Fast Company,* October/November, 1996.

Wheatley, Margaret. *Leadership and the New Science.* San Francisco: Berrett-Koehler, 1992.

Wheatley, Margaret J., and Myron Kellnor-Rogers. *A Simpler Way.* San Francisco: Berrett-Koehler, 1996.

Index